Introduction to
World Vegetation

A. S. Collinson
Principal Lecturer, Sunderland Polytechnic

London
GEORGE ALLEN & UNWIN
Boston   Sydney

First published in 1977

© George Allen & Unwin (Publishers) Ltd, 1977

ISBN 0 04 581012 5 cloth
      0 04 581013 3 paper

FOR PAT

Printed in Great Britain
in 10 on 12 point English
at The Lavenham Press Limited, Lavenham, Suffolk, England.

Preface

Over the last few decades, the interpretation of the subject-matter common to ecology and biogeography has undergone a radical transformation. The new understanding of the unity of the biosphere and the extent and scale of the processes within it make necessary not only incorporation of new data but also constant reappraisal of data and theories of long standing. This book aims to provide an introduction to a wide range of topics and problems in ecology and biogeography, especially as they affect the understanding of the spatial aspects of the world's vegetation cover. Particular emphasis is laid on references and suggestions for further reading, which have been selected to include many major review articles. To guide the reader's further study the author has added short comments where appropriate and classified into two groups those particularly recommended. References with a single asterisk are those the author feels would be particularly appropriate to expand the material of the chapter. Double asterisks indicate those which might be most appropriate as initial further reading.

The variety of vegetation dealt with in a book of this sort is necessarily so wide that, to avoid creating a misleading impression, even so-called 'type example' photographs have been omitted. There are now available many easily accessible sources of photographic illustration of vegetational appearance to which the reader can refer. Particularly recommended are those in N. Polunin, *Introduction to plant geography* (Longman, 1960); R. Good, *The geography of the flowering plants* (Longman, 1974); and the copiously illustrated elementary text, *World vegetation* by D. Riley and A. Young (Cambridge U.P., 1966). Other useful sources are indicated in the reference lists.

The book is divided into two parts, the first dealing with the principles and problems of vegetation study, the second with the modern interpretation of the nature of the world's major vegetation-types with the effects of Man placed in context.

Acknowledgements

For their many helpful comments and advice in the preparation of this book, the author would like to thank sincerely Mr P. Stott of the School of Oriental and African Studies, University of London, Mr L. H. Williams of Middlesex Polytechnic, Mr I. D. White of Portsmouth Polytechnic and Dr P. King of the Physics Department, Nottingham University.

For their kind permission in allowing the author to use their illustrative material, the author would like to thank the following individuals and organisations: Figures 39 and 52, the Association of American Geographers; Figures 30 and 38, Blackwell Scientific Publications; Figure 37, S. G. Daultrey and the Geographical Journal; Figures 21 and 44, the Ecological Society of America; Figure 19, Harper & Row Inc; Figure 22, the Ronald Press, New York; and Figure 31, Dr I. A. Rorison of Sheffield University.

Contents

PART ONE

Environmental and Ecological Principles

1

Plant geography and evolution

1.1 The development of the plant kingdom on land

In the total span of life on Earth (probably between 3000 and 3500 million years) plants have clothed the land abundantly for only some 400 million years. Their relatively late evolution indicates the difficulties presented by the land environment in terms of feeding, support and reproduction problems. One has only to compare the relative simplicity of seaweeds with the complex adaptive features of the land plants to see how many evolutionary changes in external morphology, internal anatomy and life history have been encouraged by the land environment (Figure 1). Not that all the problems were solved at once. Figure 2 indicates the successive stages that have occurred over 400 million years or so. The logic of this evolutionary story is much more apparent than real, however. From Figure 2 it can be seen that (a) there are a number of extinct groups, shown by the lines ending in a cross, and (b) that the evolutionary steps are rather halting. Thus, although there are signs of flowering plants possibly from as early as Triassic times (Axelrod 1970), their rise to dominance occurred only during the Cretaceous period some 135 million years ago. As to why they should then have rapidly 'taken over the world' is, as Darwin said, 'an abominable mystery'. Their success certainly did not eliminate the more primitive groups. The mosses, liverworts, ferns and conifers, for example, seem to have survived quite well in spite of the lack of reproductive systems as sophisticated as those of the flowers.

How then can we account for these apparent anomalies in the evolutionary story? The simple answer is that we cannot do so completely. We really know very little as to why some groups succeed while others are extinguished, nor do we know why this seems to happen irregularly (Stebbins 1974). We can certainly see in what ways favourable mutant genes become established in present populations and can say something fairly sensible about the operations of chance in

13

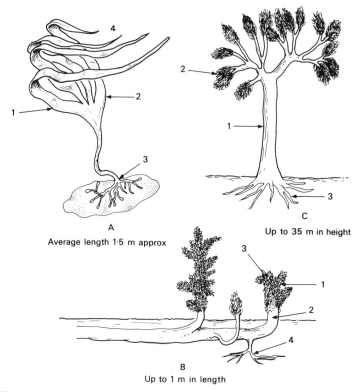

A
Average length 1·5 m approx

C
Up to 35 m in height

B
Up to 1 m in length

Figure 1. Morphological adaptations encouraged by the land habitat.
A—a modern seaweed (Oarweed): (1) simple, relatively unspecialised
form; (2) support provided by water with direct diffusion to all parts
of water, light and nutrients; (3) no true roots, instead a simple
anchorage by a 'holdfast'; (4) reproduction by free-swimming gametes
released into the surrounding water and dispersed by currents.
B—*Asteroxylon,* a simple Devonian vascular plant: (1) leaf-like
appendages for photosynthesis; (2) stem-like organ with vascular tissue
transporting water and nutrients; (3) specialised spore-bearing organs;
(4) root-like appendages with limited ability to absorb water and
nutrients. C—*Lepidodendron,* an Upper Carboniferous lycopod tree:
(1) division into parts with specialised functions; (2) branches bear
simple leaves and spore-producing organs; (3) true roots adapted to
support and absorption of water and nutrients.

Figure 2. The evolution of the major plant groups. A—primitive land
plants, mainly green algae in a few well-watered locations. B—first
forests with tree-size plants during the Devonian period developing to
luxuriant forests in the Carboniferous. C—fairly uniform forests
dominated by conifers and cycads reaching peak during the Jurassic.
D—forests dominated by flowering plants. E—appearance of
widespread grasslands.

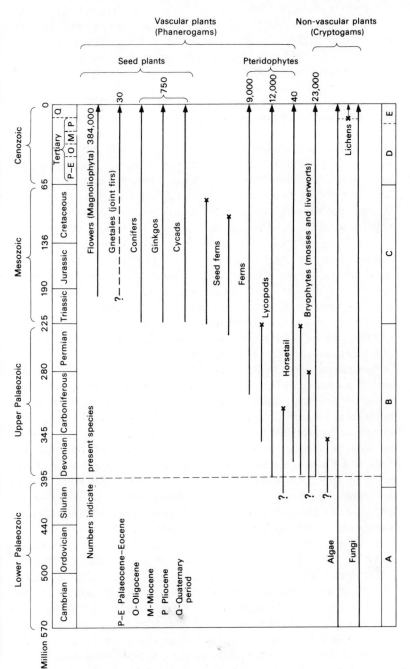

this process (Monod 1972), but, as the precise circumstances of the evolution of major groups are unknown to us, we can only speculate and theorise on the basis of limited evidence. Indeed the story of plant evolution is a very fertile field for speculation, some of it founded on very slim evidence indeed.

1.2 The evolution of geographical pattern in the world's flora

From the time when the flowering plants spread rapidly round the world to the present day, the plant cover of the world's land masses has never ceased to evolve. Mutations (changed inheritable characteristics) have emerged continuously and their survival and establishment has depended on the ecological circumstances prevailing at the time of their appearance. Climates have changed, often drastically; continents have shifted their positions, isolating some plant groups, joining others; new mountain barriers have arisen and new animal species have evolved. Thus the world today does not have a uniform flora: the thousands of species of plants are spread in a jumble of individual distribution patterns. However, it is not a completely haphazard jumble.

Major prehistoric environmental changes ('palaeoecological events'), like the splitting of a supercontinent or the advent of an ice age, tend to affect large areas simultaneously, so many species come to have greater or lesser coincidence in their distributions and we can pick out major and minor regions where there is some coherence in the patterns. We can even draw maps generalising these, as in Figure 3. Note that this is not a map of vegetation types but a map of the world's **flora.** The distinction between the two is important. Vegetation may be defined as the kind of plant cover in an area—forest, grassland, etc.—but the flora of an area is the sum total of all the plant species in it. Thus, although two floral regions (e.g. the Amazonian and West African forests) may be similar in vegetation because both are tropical forest of one kind or another, they might have little in common botanically and share only a relatively few genera (see Note 1, p. 29).

To define a floral region of the scale of those in Figure 3, the botanist relies on the geographical distribution of plant genera. These are mapped individually, and from comparison of distributions a number of natural units emerge containing suites of genera which are distinctively characteristic of particular regions. These are then grouped at a higher level into kingdoms which may have many fewer genera in common than those at a lower level. Thus Good (1974) lists only 250 genera as common to the Palaeotropical and Neotropical realms.

There are no plant types which are common to all the regions shown

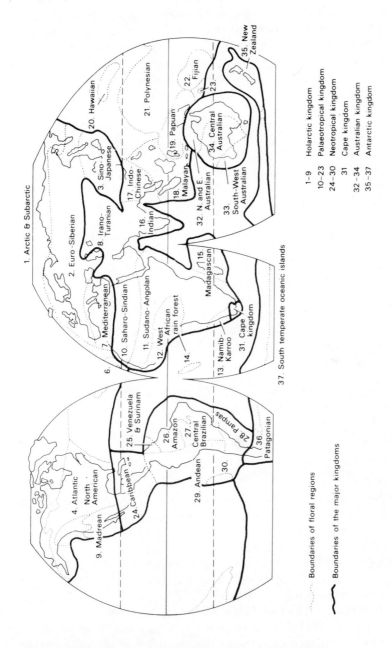

Figure 3. The world's floristic regions (after Takhtajan 1969).

in Figure 3. However, some types are very wide-ranging and these are divided into three groups: **arctic-alpine** plants, which are mainly perennial herbs; **temperate** plants, widely distributed in the wetter north temperate zone especially; and **pantropical** plants, almost universally weedy herbaceous species present throughout the tropics. Few areas are without restricted or **endemic** species, and several degrees of endemism are recognised. At one end of the scale are the broad endemics which occur throughout a major region. At the other end are the narrow endemics which are confined within very restricted environmental and geographical limits. Some plants occur in two or more regions, which may be thousands of miles apart, and are called discontinuous or **disjunct** species (Thorne 1972). To find theories which account satisfactorily for many instances of this last kind of distribution is one of the most difficult problems in biogeography.

Most biogeographical theories must take account of three main sources of evidence. First it is essential to know what a particular species needs to enable it to survive (a study known as **autecology**). Evidence generated by such a study may throw light on the conditions which might limit its spread. Secondly, the fossil record provides evidence of past distributions and what prehistoric environmental changes may have occurred to influence the patterns found today. Finally, by comparing the distributions of species for particular regions theories can be formulated which take account of correlations and discrepancies.

If all these methods told a coherent story it would be very satisfactory, but unfortunately they do not. Often, even within the same field of study, there may be serious conflicts of evidence. These conflicts are the springboard for new theory formulation and for the search for fresh evidence.

1.3 The evidence of autecology

The study of the environmental requirements of individual species in the herbarium and laboratory can often indicate the range of conditions which they can tolerate. This range is usually very different from the range of conditions which appears to limit them in nature. Plants reared in artificial environments repeatedly turn out to be surprisingly tolerant of conditions never found in their natural habitats. Why this should be so is not completely clear; it certainly implies considerably more genetic resources within the species than they usually display in their natural habitats. However, under natural conditions a species can usually be assigned an **ecological range,** i.e. a characteristic range of ecological conditions in which it is habitually found. The areal limits of this ecological range set its **geographical range.** So by identifying areas with appropriate conditions elsewhere we can assign a **potential** geographic range to a species. Probably no species fully occupies its potential geographic range and there may be several reasons for this: physical barriers of mountains and oceans; climatic barriers such as deserts and too-humid regions; poor seed mobility; and the presence of plants which already command the ecological resources required by the species. Some species appear to be able to overcome these barriers with ease, but others find even apparently feeble barriers insuperable. In the former group are many pantropical weedy species; in the latter, the members of the southern beech genus, *Nothofagus.*

Generally, all wide-ranging types contain groups of local populations which form genetically distinct races. Many studies of the chromosome content of plants, hybridisation (crossing) experiments, and the classic method of growing plants from different regions side-by-side under uniform conditions point to this. The local populations within these groups (called **ecological races** or **ecotypes**) are made up of individuals with similar environmental tolerances and are specialised to meet local environmental conditions. Their existence has been demonstrated in many woody and herbaceous species in relation to the geographical variation of light, heat, moisture, soil chemistry and so on (see Figures 21 and 30).

The presence of many ecological races in a species, combined with a wide tolerance range in individuals, results in a very broad tolerance range in the species as a whole. Thus the ability to disperse over many varied regions is enhanced. In fact, the ability to generate ecotypes seems to be a prerequisite for wide range in plants. Why some plants (e.g. the American prairie grasses) should have this ability while others seem genetically conservative and poorly fitted for wide dispersal is not well understood. Some narrow endemics may be newly evolved and may have not had time to disperse, e.g. the live oaks of California. Others—especially woody tropical species—seem to have changed little genetically for millions of years. Yet others, such as the Californian redwoods, are relict populations of species that were once much more widespread (Figure 4).

Although a good deal of the present distribution patterns can be understood on the basis of genetical study, tolerance range, and the presence of geographic barriers, these alone are insufficient to explain some of the most intriguing problems in plant geography. Without the dimension of time provided by the study of plant fossils, the evidence of present distributions is almost meaningless.

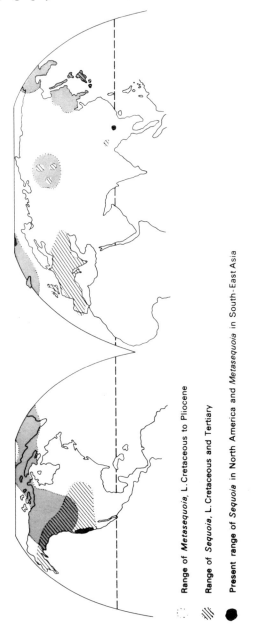

Range of *Metasequoia*, L.Cretaceous to Pliocene

Range of *Sequoia*, L.Cretaceous and Tertiary

Present range of *Sequoia* in North America and *Metasequoia* in South-East Asia

Figure 4. Past and present ranges of two Arcto-Tertiary forest genera. (From Schuster (1972), Frenzel (1968) and Takhtajan (1969).)

1.4 The fossil record of past distributions

It is no accident that the most complete fossil records of land plants are in rocks that formed in shallow, swamp-filled lakes or lagoons, nor is it surprising that there are considerable gaps in the record. Also, both macrofossils (leaves, wood, bark, seeds, etc.) and microfossils (spores, pollen) present considerable problems of interpretation and identification. For example, they may have been carried long distances from their source and may mislead in the reconstruction of the flora of a particular area. Moreover, the primary stage of evolution in flowering plants seems to have been associated with insect rather than wind pollination. Thus, little pollen may have been released and species that may have been important may hardly be represented as fossils at all. In the mid-Cretaceous Dakota flora of Kansas, for example, over 200 angiosperm (flowering plant) types can be identified by macrofossils, but flower pollen represents only 5 per cent of the total pollen types present (Axelrod 1970).

In spite of these difficulties, it is possible to conclude that, in almost every case, the fossil distribution of a species is different—often radically so—from that of today. Not only are the species distributed differently, but so are the vegetation types. Thus in the London Basin Eocene rocks there are abundant remains of a flora which has been described as tropical rain forest with coastal swamps (Chandler 1961). The fossils are well-preserved with a wealth of detail, so for once there are few problems of identification. The flora has been described as resembling the present-day vegetation of coastal Malaya. However, even these clear remains present considerable problems of interpretation. For example, were the fossil fruits of the *Nypa* palm common in these rocks fossilised *in situ,* or were they transported from possibly distant localities? There is also a considerable conflict of evidence when these rocks are compared with the Eocene rocks of the Paris Basin. Here there are thick gypsum deposits indicating arid conditions not two hundred miles from the supposed swamp forests. Today the nearest geographical equivalent might be the coast of Ecuador, but whether the peculiar conditions of that coast can be postulated for Upper Eocene Europe is another matter.

Whatever the ultimate resolution of these difficulties might be, one conclusion at least seems inescapable: that from Eocene times to the present day there has been a dramatic transformation of the climatic environment.

1.5 The evidence of climatic change

When fossil floras from around the world are compared, the evidence is unmistakable that during the last 65 million years the earth has become colder (see Turekian (1971) and Stehli (1973)). Figure 5 indicates the course of climatic change in the northern hemisphere during that time. In the southern hemisphere there is fair evidence that the glaciation of Antarctica began as early as the Eocene period.

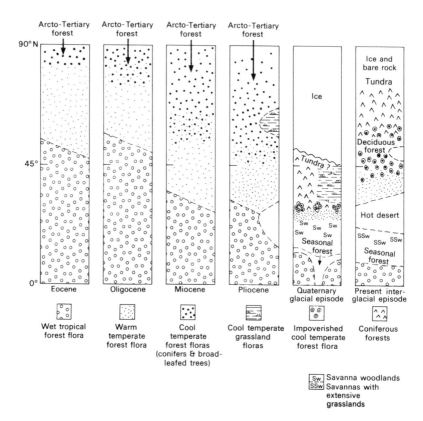

Figure 5. The course of vegetational change on the west coast of a hypothetical continent in the northern hemisphere during the Cenozoic. Note the evolution of marked latitudinal zonation of vegetation during the Quaternary.

The culmination of the process of climatic change occurred with the onset of widespread continental glaciation around two million years ago. (See Figure 6 for the pre-glacial vegetation pattern of Western Europe.) Since that time repeated cycles of glaciation have devastated the floras of the world, extinguishing many species, stimulating the evolution of adaptations in many others, disrupting or modifying the geographic range of the majority. As one eminent botanist has pointed out (Good 1974), the geography of the land plants today is the geography of a flora lately subjected to a major disaster.

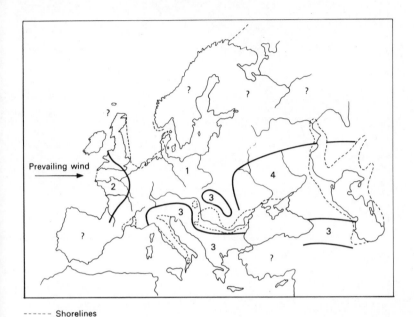

------ Shorelines

Figure 6. The vegetational pattern of Europe at the end of the Pliocene (after Frenzel (1968) and Wijmstra and Van der Hammen (see Turekian 1971)). (1) cool temperate Arcto-Tertiary forest (conifers and deciduous trees as mixed stands); (2) oceanic forest of pine, *Sequoia* and other conifers with heather undergrowth; (3) mixed warm temperate forest with warmth- and moisture-loving trees; (4) forest with prominent grasslands.

The most significant effects of glaciation were those felt in western Eurasia where the richly varied forests—the so-called Arcto-Tertiary forests—which had clothed the lands for so long in the upper Tertiary, were utterly destroyed. The broad-leaved forests were reduced to a few genera—oaks, limes, hornbeams, beeches, elms, etc.—surviving in the Balkans and Apennines (Figure 7). In North America and eastern Asia continental migration routes to the south allowed the Pliocene types to retreat much more easily before the onslaught of cold conditions, and the warmth-loving trees—the walnuts, pawpaws, sweetgums, hickories, persimmons and tulip trees—were able to shelter in the southern Appalachians and south-east China.

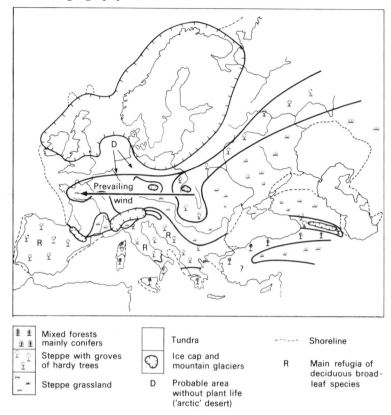

Figure 7. Vegetational pattern of Europe during the last glaciation (after Frenzel (1968) and Wijmstra and Van der Hammen (see Turekian 1971)).

In the Tropics, Quaternary climates seem to have alternated between rainy periods (pluvials) and drier spells (interpluvials). The former had the effect of destroying large areas of level erosion surfaces inherited from Pliocene times, so that new varied mosaics of different soils were created. In the latter periods the areas covered by the wet tropical forests were reduced, often to isolated regions. In West Africa, for example, the tropical forest trees in the drier interpluvials seem to have been confined to only three refuges in Liberia, western Ghana and the Cameroons. The resulting impoverishment of the flora of this region and the evolution of new plant types are still evident today.

In the highest latitudes of all in the northern hemisphere, the tundra as a vegetation type seems to have been destroyed and to have re-evolved several times during the Quaternary (Frenzel 1968).

1.6 The progress of re-colonisation in the present inter-glacial

About 8500 years ago the great continental ice sheets of Eurasia and North America began to wane and the process of re-colonisation began yet again as it had done so often throughout the previous two million years or so (see Note 2, p. 29). From the plant remains (especially pollen grains) which have built up in the peat bogs of Europe and North America since that time, the course of climatic change and re-colonisation has been constructed by many workers. One surprising fact which emerges from this work is that the time-table of changes on the two continents does not seem to be synchronous. Nor apparently is it even synchronous within North America. However, both continents are alike in that they display an 'optimum' phase of mild climate (regarded rather ominously by some geologists as the most favourable climate in the present inter-glacial) which is called 'Atlantic' in Europe and the 'Hypsithermal Interval' in North America. At this time in Britain trees grew at higher altitudes than before or since, and clad the Scottish, Welsh and English uplands to around 800 metres (2500 ft) with dense oak forest. Various other climatic phases have been recognised in Western Europe as shown in Table 1.

In North America it would seem that some sort of zonation existed throughout the last glaciation (the Wisconsin) and, as the ice retreated, the plant communities migrated irregularly in its wake, grasslands into Kansas, Nebraska, South Dakota and later Illinois, Iowa and Central Indiana, and deciduous forests into the lands from the Ohio river eastwards.

The pioneers of pollen stratigraphy assumed that changes in the pollen record usually meant changes in the climate, but over the years, in Western Europe at least, this theory has been hard to support. For example, in Britain Godwin put the Atlantic/sub-Boreal boundary (originally identified in Scandinavia) at a level where elm pollen practically disappears from the record. Diligent searching for other evidence of climatic change at this level has yielded very little. What explains the spectacular decline of one genus simultaneously over wide areas of Europe? With its present ravages, it might be suggested that Dutch Elm Disease or something similar could have been the cause, but there is another clue. Both anthropologists and archaeologists have pointed out that the foddering of animals on leaves of the elm and other plants is still practised in parts of Europe and may have been widespread amongst the herdsmen of Neolithic times. Certainly, hoards of elm leaves have been found in excavated sites. Perhaps the elm might have been cut so extensively—the easily accessible younger trees especially—that it almost became a rare plant (Pennington 1974).

Years	Climate	Vegetation	Pollen Zone
600 BC	Sub-Atlantic. Cool, wet.	Increase in birch. Elm very low, also lime. Beech and hornbeam increase.	VIII
1250 BC	Sub-Boreal. Drier, somewhat continental.	Slight decrease in alder, lime, possibly oak. Slight increase in birch.	VIIb 'Elm Decline'
4000 BC	Atlantic. Moist, warm, oceanic.	Mixed oak forest. Dominant oak with alder, elm and lime. Hazel abundant but pine and birch very reduced.	VIIa Flandrian II
(?)4900 BC	Boreal. Continental, increasingly warm.	Decreasing pine. Abundant hazel. Increasing elm and oak.	VI
(?)5600 BC		Pine dominant with birch and hazel.	V
(?)6300 BC	Pre-Boreal. Cold.	Dominant birch with pine.	IV Flandrian I

Table 1 Post-glacial climatic and vegetational change in Britain

(Dating from Renfrew (1974). Climate and pollen zones from Pennington (1974).) Pollen zones I-III are late glacial fluctuations. Note that the dates for VIIa-VIII are derived from the Arizona/Pennsylvania mean calendar table (Renfrew 1974) and may err ± 200 years. Flandrian I and II are major marine transgressions (West 1968).

Whether this is the right explanation or not, it serves to indicate that the present inter-glacial has seen the rise of man as one of the major ecological agents. A second example serves to illustrate this point: In Denmark, variations of tree pollen above the level of the elm decline are associated with episodes of clearing, burning and cultivation called 'Landnam' clearances. Experiments using stone axes newly-hafted in the manner of the Neolithic show that these were very effective means of clearing trees for such cultivation. In one experiment in western Jutland some 600 square metres of beech forest were cleared in four hours by three men (Pennington 1974).

1.7 Plant geography and continental drift

Plants had to contend not only with a deteriorating climate during the last 60 million years, but also with a slow inexorable shift of the continental masses. The theory that the Earth's crust is composed of enormous rotating rigid plates is now firmly established in the geological sciences, and other disciplines have revised their theories to take account of the new evidence. In many ways, however, it has raised more problems than it has solved (see Figure 8). On the one hand, the linking or near-linking of the north Atlantic lands allows us to say that the Arcto-Tertiary flora could have had access to sub-Polar migration routes which would account for its unified character. On the other hand, India appears to be stuck out as an isolated land mass in the middle of an ocean, but with affinities in its fossil flora and fauna with the other southern continents. Hallam (1972) has pointed out this conflict and suggested that the fossil evidence be used to refine conclusions based purely on geophysical considerations. In this case, for example, the fossil evidence suggests that India might have been much further west than is postulated. The resolution of issues like this can only await the accumulation of further data by palaeontologists and geophysicists.

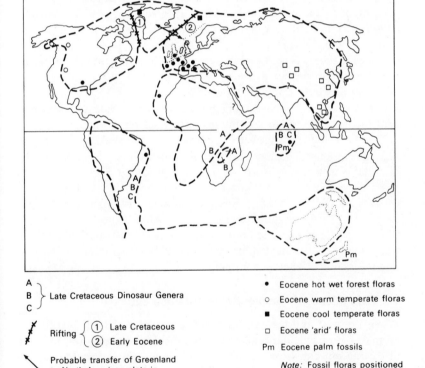

A ⎫
B ⎬ Late Cretaceous Dinosaur Genera
C ⎭

Rifting ⎰ ① Late Cretaceous
⎱ ② Early Eocene

Probable transfer of Greenland to North American plate in early Eocene

• Eocene hot wet forest floras
○ Eocene warm temperate floras
■ Eocene cool temperate floras
□ Eocene 'arid' floras
Pm Eocene palm fossils

Note: Fossil floras positioned on reconstructed continents

Figure 8. Possible position of continental masses at Late Cretaceous to Early Eocene transition (Palaeocene) (compiled from data in Dietz and Holden, Pitman and Talwani, Keast (1971), Hallam (1972) and Takhtajan (1969)). Note: the high latitudes reached by hot wet forest floras; and the anomalous positions of the Cretaceous dinosaur genera and Eocene palm floras.

Such problems have not prevented the application of continental drift theory to the interpretation of existing difficulties. One example is that old chestnut of biogeography, 'Wallace's line' (Figure 57). From its identification last century, the existence of this curious division in the faunal realms of South-East Asia has intrigued biologists and geographers. Although its precise position is variously interpreted by different authors, all are agreed that within the Indonesian archipelago a remarkably sharp division exists in its fauna and also amongst the conifers, mosses, liverworts and primitive flowering plants. No satisfactory explanation of this break has ever been given. One theory which takes advantage of the acceptability of continental drift (Schuster 1972) suggests that the line for both plants and animals is the result of the recent collision of New Guinea/Australia with South-East Asia. The concept of whole continents acting like gigantic Noah's Arks is a delightful one, but it will need a great deal more substantiating evidence before it can be accepted. As Schuster points out, 'phytogeographic theory is particularly subject to distortion that serves (a) to reinforce a previously accepted viewpoint, and (b) omits all the potentially embarrassing bits of evidence'.

Some of these 'embarrassing bits of evidence' which continental drift theory does little or nothing to explain are represented by species of limited ecological range with widely disjunct distributions. Examples include: an arctic-alpine grass (*Trisetum spicatum*) found in the mountains of Eurasia, North America, New Zealand and the Andes; some species like the creosote bush common to the semi-deserts of California, Chile and north-west Argentina; and some plant types common to the deserts of America, South Africa, the Near East and India (e.g. *Gossypium* and *Malvastrum* species). Distributions of these kinds represent a considerable challenge to biogeographical research, and it is doubtful whether continental drift will provide all the answers.

1.8 The evidence of comparative distribution patterns

When the past and present distribution patterns of both plants and animals for a particular region are assembled for comparison, the task of sorting out consistent theories to explain them is formidable. No better example could be found than that of the islands of New Zealand which have been truly described as a 'biogeographic riddle' (Keast 1971). Apart from bats, the original fauna contained no living or fossil mammals. Fossil dinosaurs are also unknown and there are no turtles or frogs. Until recently, the dominant land vertebrates were the flightless moas whose ancestors appear to have been flightless in the Early Tertiary when many marsupials already existed. Did one group walk there and the others not? If so, where was the land they crossed? The flora is equally strange. In the early Tertiary, New Zealand had a rich flora of palms, as did India, yet these are unknown in Australia. Analysed by Hooker in 1860, the then flora showed 50 per cent of species (mainly conifers) to be endemic with 25 per cent shared solely with Australia and 12 per cent with South America only. Many plant geographers have argued that this flora indicates (a) long isolation for New Zealand, and (b) close connections at some time between New Zealand, Australia, Antarctica and South America. Both Darwin and Wallace argued against this and suggested that the disjunct distributions of southern plants were relics of once widespread populations which had become extinct elsewhere. Darlington (1965) argues that the Antarctic continent formed the connecting link and that migration routes existed until Miocene times at least. Yet New Zealand acquired neither dinosaurs nor mammals and Australia no palms. Moreover, the evidence of early Tertiary glaciation in Antarctica is quite strong. So the puzzle remains.

New Zealand is, admittedly, an extreme example, but it demonstrates the kind of study that biogeography is. More than 100 years of patient research have served only to raise more questions. It is precisely this quality of patient detective work that makes the subject so attractive.

Notes

(1) Page 16. Set out below is an example of botanical classification:
   Dog rose—*Rosa canina* (Linnaeus); *Rosa styla* (Bastard);
   Kingdom: Planta;
   Subkingdom: Phanerogamae or Embryobionta (vascular plants);
   Division: Magnoliophyta (also called Angiospermidae—flowering plants);
   Class: Magnoliatae (also called dicotyledonous flowers or dicots);
   Subclass: Rosidae;
   Order: Rosales;
   Suborder: Rosinae;
   Family: Rosaceae;
   Subfamily: Rosoideae;
   Tribe: Roseae;
   Genus: Rosa;
   Species: *Rosa canina* (L) or *Rosa styla* (Bast.)

(2) Page 25. Direct dating of the last few tens of thousands of years has for a long time been made from the radioactive carbon (^{14}C) content of organic material. From the known rate of decay of ^{14}C—its 'half-life'—it was accepted

that an estimation of the amount remaining *in situ* would enable the amount lost, and therefore the age of the material, to be calculated. However, work on the long-lived bristle-cone pine of California and the redwoods, indicates that the ^{14}C content of their growth rings has been *variable* over the (literally) thousands of years which these plants can survive. From these variations the previously accepted dating of the recent past has undergone considerable revision in recent years and it appears that earlier accepted datings (especially for the period before 600 BC) are serious underestimates.

References and Further Reading

Evolutionary theory

Axelrod, D. I. 1970. Mesozoic palaeogeography and early angiosperm history. *Bot. Rev.* **36,** 277-319.

**Delevoryas, T. 1964. *Plant diversification.* New York: Holt, Rinehart & Winston. (A clear account of the main stages of plant evolution.)

Monod, J. 1972. *Chance and necessity.* New York: Knopf; London: Collins.

Stebbins, G. L. 1974. Adaptive shifts and evolutionary novelty: a compositional approach. In Ayala, F. J. & T. D. Dobzhansky (eds), *Studies in the philosophy of biology,* 285-306. London: Macmillan. (This book is the result of a private symposium on biological problems by some of the foremost thinkers in the field.)

Takhtajan, A. 1969. *Flowering plants: their origin and dispersal.* Revised English language edn, trns. C. Jeffrey. Edinburgh: Oliver & Boyd. (Often speculative, but a clear account of the evolution of this group.)

Plant geography

Darlington, P. J. 1965. *Biogeography of the southern end of the world.* London: Oxford University Press.

*Good, R. 1974. *The geography of the flowering plants,* 4th edn. London: Longman. (The standard reference in English on plant geography.)

Keast, D. 1971. Continental drift and the biota of the southern continents. *Qu. Rev. Biol.* **46**(4), 335-78.

*Polunin, N. 1960. *Introduction to plant geography and some related sciences.* London: Longman. (Somewhat outdated but with useful illustrations.)

Schuster, R. 1972. Continental movements, Wallace's Line and Indomalaysian Australian land plants: some eclectic concepts. *Bot. Rev.* **38**(1), 3-87.

Thorne, R. F. 1972. Major disjunctions in the geographic range of seed plants. *Qu. Rev. Biol.* **47**(4), 365-411.

Geological evidence

Chandler, M. E. J. 1961-6. *The lower Tertiary of southern England,* Vols I-IV. London: British Museum (Nat. Hist.).

*Frenzel, B. 1968. The Pleistocene vegetation of northern Eurasia. *Science* **161**(3842), 637-48. (Excellent maps.)

*Hallam, A. 1972. Continental drift and the fossil record. *Sci. Am.* **227**(5), 56-70.

Renfrew, C. 1974. Changing configurations. In Renfrew, C. (ed.) *British prehistory.* London: Duckworth. (Especially useful summary of carbon dating.)

Stehli, F. G. 1973. Review of palaeoclimate and continental drift. *Earth Sci. Rev.* **9**(1), 1-19.

Pleistocene changes

**Pennington, W. A. 1974. *The history of British vegetation,* 2nd edn. London: English U. Press. (A very clear account of post-Pleistocene changes.)

*Turekian, K. K. (ed.) 1971. *Late Cenozoic glacial ages.* New Haven: Yale U. Press. (See particularly the article by Van der Hammen *et al.* on the Late Cenozoic of Europe.)

*West, R. G. 1968. *Pleistocene geology and biology.* London: Longman. (A very thorough guide to a complex period of Earth history.)

2

Ecology, biogeography and energy variations

2.1 The principles of ecology

To answer two questions which concern the ecologist and biogeographer, 'Where do species live?' and 'Why there?', geological history can only take us so far along the road. We need also to know the way plants live now and what in their present environments encourages their growth or limits their spread. It is to the science of ecology that we must look for insights, techniques and principles to guide our understanding in this task.

Ecology has recently caught the public attention in no uncertain way and has come to the fore in the last two decades as much as a result of a change in its own methods as of a desire on the part of the public to learn some scientific natural history.

The development of the dynamic and rapidly growing study of today can be dated almost exactly to the publication in 1942 of a paper by a brilliant young American biologist, F. E. Lindemann—later a tragic victim of World War II—concerning the quantification of the relations between organism and environment, especially the exchanges which take place in energy and materials. From that time the subject of ecology has been transformed, by the application of measurement and mathematics, into a considerably more precise scientific study. With the publication of each paper since that time on matter and energy exchanges it has become more and more apparent that organism and environment are so inextricably intertwined that to consider them separately is almost impossible. Only a concept of ecology which places the unity of living and non-living matter at its heart can do full justice to natural phenomena on this planet.

This insight is one of the most important contributions science has made to the understanding of our world so far this century, and its echoes reverberate into fields far from the ecologist's laboratory.

The key to modern ecology lies in the concept of the **ecosystem.** This concept implies that (a) all plant and animal communities without exception, together with the non-living world they inhabit and partially fashion, form unified systems, and (b) these systems are maintained by the production and flow of energy within them and the circulation of materials between the living and non-living matter which composes them, i.e. they are wholly unified (**holocoenotic**). Ecosystems can be as small as a living-room fish tank or as large as an equatorial forest, because the principles which govern their operation remain the same whatever the scale. It is implicit in this concept that there can be no duality of 'living things' on the one hand and 'environment' on the other. Plants and animals are as much environment as the air they breathe or the soil they live in or the food they eat; the duality is an illusion.

It must always be remembered, however, that each plant and animal is genetically distinct and each must work out its own relationship with whatever circumstances prevail in any particular ecosystem. If its genetics do not 'fit' it will not become established; if they do then under the right circumstances it may well become established.

Following Odum (1963) the circumstances which shape the character of ecosystems can be listed as:

(a) energy, its production and flow;
(b) materials—the inorganic chemicals (**nutrients**) out of which living matter is formed;
(c) conditions—the prevailing circumstances of place in which the ecosystem operates; and
(d) community—the plants and animals and their interactions which compose the living part of the ecosystem.

In addition, the activities of Man play a greater or lesser role in most ecosystems, although in some respects (e.g. radioactive pollutants) his effects are virtually universal. A practical ecologist would worry very little about the niceties of whether a particular factor represented 'condition' or 'community' or 'material'. The categories are used only as a broad organisational device through which the component parts of the ecosystem can be isolated for examination. One thing that can be said about these determinants of ecosystem operation is that each has a very definite geographical aspect at almost every scale from that of the Earth biosphere (i.e. the atmosphere, oceans and rocks containing living material) to the strictly local.

Although each ecosystem is unique to some extent, broad generalisations may be possible about some of them, as certain factors such as climate or mineral supply may dominate the working of the system and may therefore have acted as a strong selective force in the evolution of the organisms concerned. Thus, similar adaptations may appear in similar circumstances. It is for this reason that broadly similar categories of vegetation—forests, grasslands, desert scrub, etc.—have been fashioned by convergent evolution from the very different groups of plant types shown in Figure 2.

2.2 Ecosystems and energy

Apart from certain micro-organisms (termed **chemautotrophs**)—for example, chemosynthetic bacteria—which obtain their energy via the oxidation or reduction of simple inorganic chemicals, the foundation of life is the sun's radiant energy. This is received by plants in three distinct forms, thus:

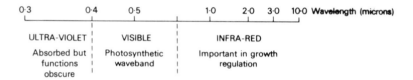

Plants absorb strongly throughout this range—apart from a very narrow band in the infra-red—but the destination of much of this radiation (especially in ultra-violet) is still something of a mystery. The quanta (or 'energy packets') of which radiant energy consists have insufficient energy in the visible light wavebands to bring about by themselves the combination of carbon dioxide and water to form carbohydrates. Instead the energy absorbed by chlorophyll—especially in the blue and red frequencies—and other pigments is transferred to electrons to produce strong oxidising and reducing agents. It is these which help to build the energy bonds in the primary products of photosynthesis, i.e. sugars and a compound called adenosine triphosphate (**ATP**). The latter acts as an energy donor to a vast number of metabolic reactions.

The amount of food created by green plants in an ecosystem is called its **gross production** and this is used by the plants themselves in their own respiration to sustain their life, but in all ecosystems they produce a surplus which is called the **net primary production.** It is this surplus which is used for growth and by the herbivorous animals and the rest of the food chain as shown in Figure 9.

At every stage in the food chain from green plant to herbivore to carnivore there is a loss of the total energy passing along the chain of the order of nine-tenths (this is the so-called '10 per cent rule').

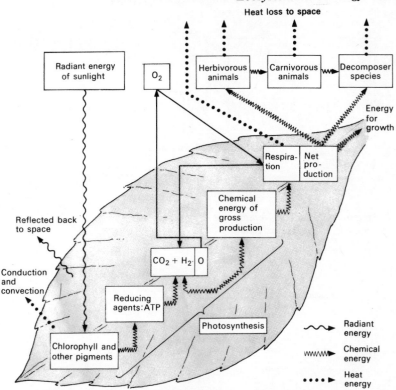

Figure 9. Energy flow in an ecosystem. Note that the units of measurement of production are normally given as the dry weights of organic matter per unit area of ground per unit of time. These may be tonnes/ha^{-1}/an^{-1}, kg/ha^{-1}/an^{-1}, or gm/m^2/an, and in some continental literature the *centner* (50 kg) may be used.

Agricultural ecosystems minimise the loss from stage to stage, and as much as 25 per cent of the total energy available as food may pass from crops to man, his animals and various pests (Figures 10 and 11).

Only a small fraction of the energy available to green plants is actually converted to food. Figures vary from one vegetation type to another. At the greatest efficiency, for example in rapidly growing barley, as much as 14 per cent of the total incident light has been observed to convert to food for short periods. However, on a year-round basis the figures are much less. Tropical crops on a year-round basis without irrigation can convert 0.5-1 per cent of the total available radiation. With irrigation, about 2 per cent of this radiation can be captured. Figures for natural vegetation are usually much lower than these; in mature vegetation perhaps below 0.5 per cent (Leach 1975).

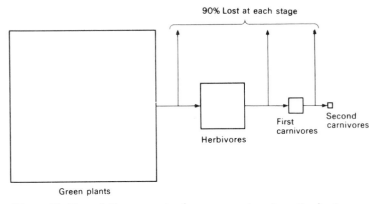

Figure 10. The relative amounts of energy passing along the food chain in a natural ecosystem.

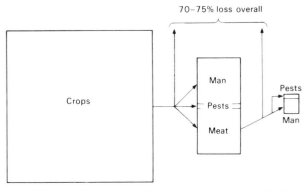

Figure 11. The relative amounts of energy passing along the food chain of an agricultural ecosystem.

2.3 Variations of production

By careful estimation of the solar radiation input to the Earth's surface and assuming that plants are fully using the fraction of radiation needed for photosynthesis, it is possible to obtain estimates of the geographical variation of potential net photosynthesis in regions where plant growth is not limited by lack of water or permanently low temperatures. Figure 12 indicates the latitudinal variation of these estimates. (Note the surprisingly high estimates for the summer months produced by increasing day length in higher latitudes.) Publication of the results of the International Biological Programme's efforts to measure actual net primary production in all the world's major

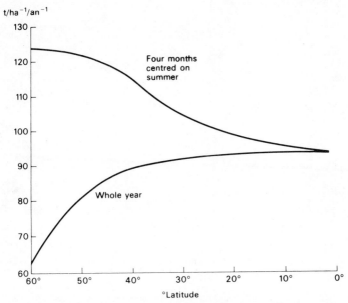

Figure 12. The latitudinal variation of potential net production rates in well-watered vegetation.

ecosystems (Lieth and Whittaker 1975) shows quite clearly that none of them even nearly approaches the theoretical limits. Only in certain tropical agricultural systems—sugar cane cultivation especially—are the theoretical and actual net production yields approximately equal (Figure 13).

The reasons why the actual and potential figures diverge so far in nature vary from ecosystem to ecosystem, but all of them can be summed up in the modern version of the agronomist Liebig's 'Law of the Minimum' which states that rate of growth is dependent on whatever factor in the environment is present in the *minimum quantities in terms of need and growth* (Odum 1963). There are *no* environments in which some limiting factor or set of factors is not present. It may be a case of excessive shading, as in the dense rain forests, too little or too much rain, some mineral deficiency or any of the multitude of interacting components which make up an ecosystem.

Man's achievement since the invention of agriculture has been to trap an increasing proportion of the sun's energy by eliminating or minimising the effects of this law. He has done this by improving agricultural techniques and by evolving new breeds of plant with growth optima better adapted to lessen the checks which nature puts between actual and potential yields.

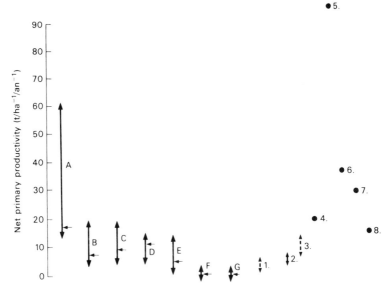

Figure 13. The variation of net primary production in various
vegetation types. A—tropical rain forest; B—tropical savannas;
C—boreal and montane coniferous forest; D—deciduous forest of the
northern hemisphere; E—steppe grassland; F—tundra; G—arid lands
(true deserts (see p. 150) not recorded). The horizontal arrows indicate
mean net primary production levels. (1) lowland wet rice farming,
Ceylon; (2) lowland wet rice, Bangladesh; (3) hybrid corn, USA; (4)
lowland wet rice using 'high-yield' varieties, Vietnam; (5) sugar cane,
Java; (6) oil palm, Congo Republic; (7) grassland, North Island, New
Zealand; (8) Scots pine plantation, England. (Figures from various
sources including Lieth and Whittaker (1975), Lemée (1967), and
Leach (1975).)

2.4 Plant response to the geographical variation of light

The seasonal and daily variations of solar radiation are a function of
latitude and are of great biological consequence. Besides its effects on
photosynthesis, other physiological activities vary in relation to the
period of illumination, and these variations (**photoperiodic responses**)
have been widely studied experimentally in relation to flowering,
germination and growth.

In middle and high latitudes most flowering plants can be classed
into long-day, short-day and indifferent plants according to whether
they flower in response to critical day or night length. Long-day plants
flower only in the late spring or early summer—the radish, iris, red
clover, spinach, the smaller cereals and timothy grass are examples.

Short-day plants flower in the late summer and autumn—tobacco, goldenrod, aster, dahlia, ragwort and chrysanthemum exemplify this group.

The dependence on day length means that long-day plants are excluded from low latitudes and that short-day plants may be excluded from higher latitudes as they may not complete their reproductive cycle before the first killing frosts arrive.

The selective force of day length is very powerful, and within particular species there may be ecological races adapted to different day lengths. In a classic piece of work, McMillan (1959) showed that the northern prairies of North America are dominated by long-day ecotypes which can take best advantage of the summer season, whilst the southern prairies are dominated by short-day or indifferent ecotypes. This goes far to solve the puzzle of how such a great span of latitude and therefore climate can be dominated by so few species.

For many years it was assumed that low-latitude plants would show little or no response to photoperiod variations. However, experimental evidence shows that they are, in fact, very sensitive indeed, possibly even more so than temperate plants. Thus over fourteen tropical forest tree species have been demonstrated experimentally to grow faster under long-day conditions, some with growth rate increments of over 200 per cent. For some varieties of rice, claims have been made that differences of as little as *five minutes* in day length are detectable (Longman and Jenik 1974). The photoperiod response in tropical plants has also been shown to interact in a complex way with temperature changes, to which species also appear to be extremely sensitive.

At more local levels the striking response which plants make to shade is of great importance in determining the structure of land plant ecosystems. Wherever a substantial cover of vegetation exists, the plants involved are subject to heavy shading for at least some parts of their lives and there can be few which do not react to this factor. Most are either shade-tolerant or shade-intolerant. In the natural development of temperate forests particularly, tolerance to light is usually a controlling factor in the progress of colonisation and the establishment of species. It has been shown that in the earliest stages of forest growth it is the light-tolerant, high-producer species which become established on open sites. As their shade develops, it is the slower-growing, shade-tolerant species whose seedlings are the most successful (Figure 14(A)).

In the humid tropical lands, the structure of mature forests is closely related to demands for light. At the upper level are the emergents, which have evolved to take maximum advantage of the light available.

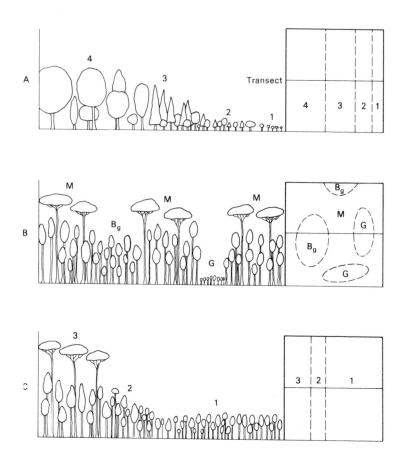

Figure 14. Three schematic models of successional patterns. A—model in middle latitude temperate deciduous forest. This example was observed on old fields in Virginia. Phases 1-4 are from (1) high-producer species demanding much light to (4) low-producer, shade-tolerant species; B—regeneration pattern in tropical rain forest. In this model small gaps are quickly occupied from surrounding mature forest (M) to give a mosaic pattern of forest overall with gap phases (G) and intermediate building phases (Bg) recognisable in the areal pattern. (If the gap is too small, species needing most light may not regenerate.); C—regeneration into a larger area in tropical rain forest. Here succession from pioneer colonisers (which may include short-lived softwoods) with (1) light-demanding species to (3) mature forest may be recognised.

In the middle and lower layers there is intense competition for light, and the leaf mosaic, spread and arrangement of the leaf blades are usually closely related to available light. At the lowest levels are shade-tolerant shrubs and herbs, slow-growing saplings and seedlings (see Figure 41). The replacement of such forest with its many shade-loving species is a more complex process than the relatively simple model outlined above for the temperate forests. Thus it has been noted that although there may be fully-mature emergent species above, their seedlings and saplings are often absent below. These light-demanding species may not even regenerate when a gap opens in the canopy if it is too small, although in larger gaps covering several square kilometres a succession from pioneer colonisers to mature forest can be recognised (Figure 14(B) and (C)).

The radiation to which plants respond by varying their growth rate is not necessarily that which they absorb for photosynthesis. There is good evidence (Bonner 1966) that growth rate variations are considerably influenced by the balance of infra-red radiation received. This seems to be absorbed by a molecule called **phytochrome** and it is this that controls the growth response so that a plant reaches a suitable size in relation to the light conditions which suit it.

2.5 Energy microclimates

Plants participate in many more exchanges of energy with their environments than simply the absorption of radiation for photosynthesis and growth. They reflect radiation—some 10-15 per cent of visible light in the case of a green field, 5-10 per cent from a coniferous forest. They transpire water vapour, and heat flows to and from them by conduction and convection. Physicists have suggested (Gates 1968) that to describe the true coupling of a plant to its environment the energy microclimate (both input and output of energy) of the leaf must be accurately measured. Moreover, they have presented mathematical formulae for doing this and have suggested that this measurement of energy inflow and outflow could be extrapolated to whole forests.

As may be imagined, the complexities of such a task, desirable though it may be, have not encouraged many workers to accept the challenge. Moreover, plant growth and behaviour can be correlated with other energy-related factors in the environment such as temperature, evaporation, and light variations. These can be applied at both the microclimatic and macroclimatic levels and have the advantage of being easier to measure. Thus the familiar climatic statistics, however suspect in the eyes of a physicist, still have relevance in describing the energy microclimate of plants. Moreover, their usefulness has been considerably enhanced by the development of models whereby, using standard environmental data, productivity patterns can be predicted (Lieth and Whittaker 1975).

2.6 Energy, land plants and geomorphology

The amounts of energy which flow to and from the habitat of a community and the individual plants which compose it, and from one community to another community, in time considerably modify the nature of the habitat itself. A soil which develops on a bare rock surface is a material very different from its parent rock and one which is in dynamic relationship with the organisms it supports. Energy is supplied to it by leaf fall and dead plant and animal material. Its chemical and physical structure is dependent on that energy supply being maintained. Once it is cut off, its nature radically alters.

The fixed energy in the form of the plant **biomass** (the weight of organisms involved in an ecosystem—see Glossary) also alters the habitat near the ground. It controls the way in which rain is received and distributed back to the atmosphere and run off, and it modifies wind speed and controls insolation and radiation. Consequently many geomorphological processes operate very differently from the ways in which they affect unvegetated surfaces. One has only to consider the vastly accelerated erosion rates consequent on deforestation or the ploughing of some grasslands to realise the erosional energy which plants divert by their presence.

Until recently the precise effects of the energy exchanges within ecosystems on the processes of earth sculpture were not well understood. That these effects are profound cannot be doubted. It might not be too sweeping a case to argue that, as the activities of green plants have largely determined the atmosphere's present composition and strongly influenced the distribution of energy within it, then precipitation and run-off must, as agents of earth sculpture, be seen as functions of the biosphere (see p. 33).

The system of energy flow briefly outlined in this chapter has been going on for 2500-3500 million years. The exchange of gases and materials facilitated by it has enabled plants to create of this planet's oceans, atmosphere and land surfaces a world almost literally fashioned in their own image, a 'meadow in the sky' as one poet has put it. The results of the recent exploration of the solar system have brought home to us how different is the Earth from its dead neighbours and how much of this difference in its outer biospheric skin is due to the dynamic energy flux which is life (Siever 1975).

References and Further Reading

General

Bainbridge, R., G. C. Evans and D. Rackham 1966. *Light as an ecological factor.* London and Edinburgh: Blackwell.

**Daubenmire, R. F. 1976. *Plants and the environment,* 3rd edn. New York: Wiley.

**Gates, D. M. 1962. *Energy exchange in the biosphere.* New York: Harper & Row. (An excellent introduction to the physicist's approach to biology and energy.)

Lindeman, R. L. 1942. The trophic-dynamic aspects of ecology. *Ecology* **23,** 399-418. (To read this paper is to understand how far our understanding has progressed since it was published.)

*Odum, E. P. 1963. *Ecology.* New York: Holt, Rinehart & Winston. (A very clear introduction to the trophic-dynamic aspect of ecosystems.)

Siever, R. 1975. The Earth. *Sci. Am.* **233(3), 82-92. (Articles from this source are available as offprints and are collected into books of readings, e.g. *The biosphere.* San Francisco: W. H. Freeman.)

Whittaker, R. H. 1969. New concepts in the kingdom of organisms. *Science* **163,** 150-160.

Ecology of plant production

Black, C. C. 1971. Ecological implications of dividing plants into groups with distinctive photosynthetic capacities. In Cragg, J. B. (ed.) *Advances in ecological research.* New York: Academic Press.

Eckardt, F. E. (ed.) 1968. *Functioning of terrestrial ecosystems at the primary production level.* Paris: Unesco.

Lieth, H. and R. H. Whittaker (eds) 1975. *Primary productivity of the biosphere.* Berlin: Springer Verlage. (Essential source of information on methods of assessing primary production. But see also p. 120.)

*Leach, G. 1975. *Energy and food production.* London: Int. Inst. for Environment and Development. (A useful summary of the energy basis of agriculture.)

Miller, P. C. 1972. Bioclimate, leaf temperature and primary production in Red Mangrove canopies in southern Florida. *Ecol.* **53**(1), 22-45.

Monteith, J. L. 1972. Solar radiation and productivity in tropical ecosystems. *J. Appl. Ecol.* **9,** 747-66.

Werner, A., W. Terjung and L. S-F. Stella 1973. Energy budgets and photosynthesis of canopy leaves. *Ann. Assoc. Am. Geogs* **63**(1), 109-30.

Energy bioclimate

Budyko, M. I. 1958. *The heat balance of the Earth's surface.* English language edn, trns. N. A. Stepanova. US Dept. Commerce: US Weather Bureau.

Gates, D. M. 1965. Energy, plants and ecology. *Ecol.* **46,** 1-13.

*Gates, D. M. 1968. Towards understanding ecosystems. In Craggs, J. B. (ed.) *Advances in ecological research.* New York: Academic Press.

Effects of light variations

Bonner, B. 1966. Phytochrome and the red, far-red systems. In Jensen, I. and K. Kavaljian (eds) *Plant biology today.* New York and London: Macmillan.

*Longman, K. A. and J. Jenik 1974. *Tropical forest and its environment.* London: Longman. (Good data on the effects of day length on tropical plants.)

*McMillan, C. 1959. The role of ecotypic variation in the distribution of the central grasslands of North America. *Ecol. Mono.* **9,** 285-308.

*Whitmore, T. 1975. *The tropical rain forest of the Far East.* London: Oxford U. Press. (Clear account of replacement and regeneration of tropical forest in relation to light factors.)

3

Nutrients and nutrient cycles

Energy, as we have seen, flows in only one direction through an ecosystem. In contrast, the materials from which living matter is fashioned are in constant circulation back and forth between plants and animals and the non-living environment.

3.1 Biogeochemical cycles

All the materials are simple inorganic elements and compounds (biogenic salts) derived from rocks, water and the atmosphere. Once involved in an ecosystem these nutrients follow pathways of greater or lesser complexity called biogeochemical cycles, set out schematically in Figure 15. Apart from water and the gases, only small fractions of the total nutrients in circulation ever escape once they are part of a stable ecosystem.

Figure 15 indicates that nutrient cycles are of two types, atmospheric and sedimentary. The two types may be illustrated by comparing briefly the nitrogen and phosphorus cycles. Both elements are needed by plants and animals in relatively large quantities, nitrogen being essential to the formation of amino acids (the building blocks of proteins) and phosphorus to the energy metabolism of living matter. Like all nutrients, whether needed in large quantities (macro-nutrients) or as traces (micro-nutrients), their availability to green plants depends on a host of micro-organisms and animals, the circulation of the oceans and atmosphere, and global and local geological processes. The various stages involved for these two elements are set out in Figure 16.

Because of the importance of microbes in mineral element circulation, the climatic and other conditions which determine the circumstances for growth of micro-organisms will have important effects on the nutrition of higher plants. Thus, in cold, wet soils the bacterial populations will be low and the soil deficient in mineral nitrogen. Organic matter may also accumulate because of its slow destruction. It is on soils like these in Britain that the bizarre plants like sundew (*Drosera* spp.) and butterwort (*Pinguicula* spp.) occur, obtaining their

45

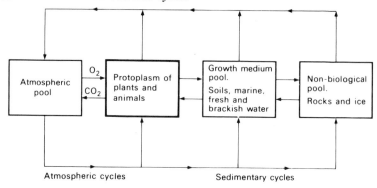

Figure 15. Simplified model of nutrient cycles.

nitrogen from the digestion of insects. In warmer climates, on the other hand, humidity favours both the fixing of mineral nitrogen and its rapid release from organic compounds. Where soil acidity is high, the rates of both humus formation and mineralisation are usually poor, whilst alkaline soils discourage mineralisation but encourage humification.

Climatic and soil-forming processes may also have important effects on a mineral like phosphorus. Free phosphates are very likely to become tightly bound to clay minerals, especially kaolinite. As this last mineral is usually an abundant product of the rock/soil weathering process in the wet tropics, there may be little phosphorus available to plants from the soil. Most of what is available is usually locked into the living biomass and re-absorbed rapidly on its release by decay before it can be bound into the clay minerals. Thus, although nitrogen and phosphorus circulate via different pathways, they are available to land plants only through the medium of the soil, and this is true of all the other nutrients except carbon and oxygen.

Although the full scale at which biogeochemical cycles operate is global and their time-scale geological, within a short time Man has had striking effects on both atmospheric and sedimentary cycles. We have added measurably to the atmospheric cycle by burning fossil fuels, and we have increased the rate of circulation of the sedimentary cycle by deforestation, agriculture and mining. Combustion and industrial chemical processing release to the atmosphere vast quantities of carbon dioxide, carbon monoxide, sulphur dioxide, hydrogen sulphide, nitrogenous compounds and hydrocarbons. Most of these, except sulphur dioxide and hydrocarbons, are released in smaller quantities than from natural sources. However, even where the natural source is enormous the effects of pollution may be detectable. Thus, it has been

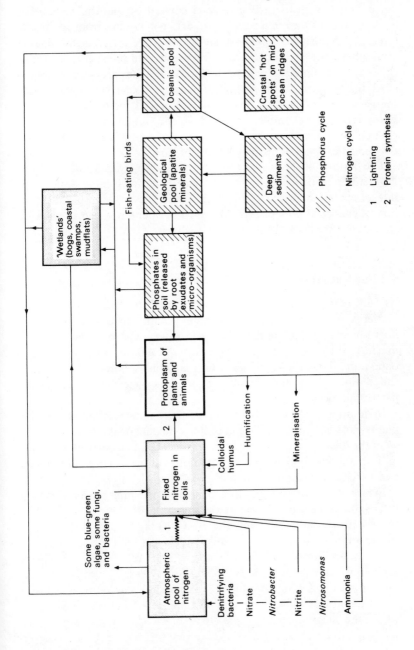

Figure 16. Nitrogen and phosphorus cycles compared. Note the importance of micro-organisms in both.

noted that the mean carbon dioxide content of the atmosphere has risen in recent times to around 320 parts per million from its former levels of 315 parts per million or so. As CO_2 is involved in the Earth's atmospheric heat balance (because it absorbs and re-radiates heat) this rise may have climatic effects in the future. In many areas deforestation is adding nutrients to the geological pools by erosion. Once released to the geological pools these nutrients will be lost for millions of years. Phosphorus, for example, attached to clay minerals swept away in rivers has to be replaced by the mining of phosphatic rock taken from the geological pool. This may be lost in turn.

The nature of Man's use of nutrients is, in fact, qualitatively different from their natural circulation. It is much more akin to the flow of energy in an ecosystem in that it is one-way and parallels the flow of energy. As the energy-flow in Man-oriented systems increases, so does the through-put of materials with it.

3.2 Nutrient groups

As indicated above, the quantities of nutrients required by living matter serve to distinguish two main groups: macro-nutrients and micro-nutrients. The former comprises water, oxygen, carbon dioxide, mineral nitrogen, sulphur, phosphorus, calcium, magnesium and potassium. The latter includes iron, manganese, boron, copper, zinc, vanadium, cobalt and possibly many other elements required by particular species.

In addition to these useful minerals, there may be many others that plants can absorb which have poisonous (toxic) effects. In fact, almost *all* of the otherwise useful nutrients may be toxic to a particular species, or, if they are present in excessive quantities or in certain combinations, to a variety of species.

Figure 17 indicates the total amounts of mineral nutrients in litter-fall circulation in some of the major vegetation-types for which figures are available. Probably the most striking feature of the chart is the enormous quantities of mineral elements in circulation in the tropical rain forests. This is not solely a direct relationship to mineral availability (due to the rapid rate of nutrient release by weathering) and the opportunity for high production, although this accounts for most of the difference. It is also a function of the relative lengths of time during which the biogeochemical cycles have been operating comparatively undisturbed on these surfaces. As we saw in Chapter 1, almost every other vegetation-type, with the possible exception of the sub-tropical forests, has suffered considerable disturbance during the last two

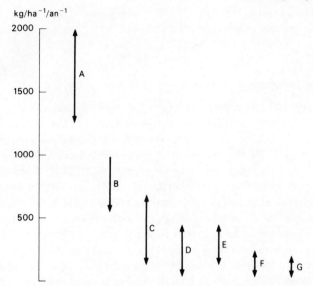

Figure 17. Range of mineral element quantities in litter-fall circulation
in various vegetation types (data from Rodin and Bazilevich 1967).
Letters in brackets indicate the predominant element in the litter.
A—tropical rain forest (Si 50-60% of leaf-fall ash); B—sub-tropical
seasonal forest (Si plus N in some); C—mid-latitude steppes (Si);
D—desert communities (Cl in solonchak soils; Si in semi-shrub desert
with annuals); E—mid-latitude deciduous forest (Ca); F—coniferous
forest (Ca—see Figure 53); G—tundra communities (N).

million years so that mineral cycles in many higher latitude environments have been in operation for only a short time. If it were possible
for us to measure the mineral cycles of the Arcto-Tertiary forests they
would almost certainly be on a greater scale than those of their
impoverished descendants today.

Figure 17 also indicates that the predominant minerals shed with the
litter vary widely from one type of vegetation to another. Some of these
minerals, for example silicon and chlorine, may in most cases hardly be
'nutrients' in any important sense at all and have little more than
nuisance value to the plants which pass them back and forth. (There is
some uncertainty about nutrients of this kind, however. For example,
silicon has been shown to be important in the metabolism of rice.)

The reason why plants cycle large quantities of minerals of possibly
limited utility lies in the way they absorb minerals from the soil. Most
plants are fairly indiscriminate in their mineral nutrition, taking
elements from the environment more or less in the proportions in which

they exist in soil water. It is for this reason that plants can easily be contaminated by man-made pollutants like toxic metals and radioactive fall-out.

3.3 The geographical variability of nutrients

The variation from place to place of mineral and other nutrient supplies has been one of the most potent factors in plant evolution, and many of the major and minor varieties of vegetation type can be ascribed to it. By the evolution of species which can patiently hoard scarce minerals, plant life can overcome many of the obstacles to production in habitats where certain minerals are hard to obtain. By their tolerance of what to other species may be toxic minerals, certain species may gain enormous territorial advantages. For example, the genus *Fouquieria* (the 'gypsum flower') in the arid zone of northern Mexico can dominate large areas of gypsum soils by its tolerance of SO_4 ions. Indeed its growth is limited to such media. However, by far the most potent factor in plant evolution in relation to nutrient supply is that of water. Hoard it as they may, plants need to breathe and photosynthesise and this implies water loss by evaporation at some time or other.

Unlike many of the other materials which can be circulated almost indefinitely in a well-established ecosystem, the renewal of water depends on the global circulation of the atmosphere and the oceans. Driven by the solar energy budget of the earth, it is outside the control of the individual species composing the individual ecosystems. In middle and low latitudes, the energy conditions of the atmosphere are such that low relative humidities and high evaporation rates may be expected at some time in almost every habitat including the tropical rain forests. Thus, water supply is a major determinant of the individual characteristics of species and the geographical pattern of vegetation. Even in high latitudes with plentiful water and low evaporation rates, plants experience difficulties of supply during the year owing to the slower rate of flow of water and the sluggish behaviour of protoplasm at low temperatures. Water supply is of such fundamental importance to understanding the geography of plant life that it will be dealt with at greater length than the other nutrients.

3.4 Water supply, plant metabolism and geographic pattern

Water is the medium in which the hydrated organic molecules of protoplasm are bathed; it participates in numerous chemical reactions, not least photosynthesis; it acts as a solvent and transporting medium;

and it maintains the form of non-woody tissues. Notwithstanding all these vital functions in plants, most of the water which is absorbed is never used. In maize, for example, 98 per cent of the water absorbed is lost and only a minute fraction, some 0.2 per cent, is used for photosynthesis. These proportions are different for different plants, but it is nevertheless true that whenever a plant is absorbing water almost the same amount is being lost from the leaves and other aerial parts by transpiration. Yet transpiration is, in fact, the physical process upon which plants rely for the absorption of water.

The process of transpiration has been described as a necessary evil because, as plants must exchange gases with atmosphere and as these must be dissolved before they can reach living matter, some water must inevitably escape to the atmosphere. In their evolution the land plants have turned this imposition to advantage by adopting it to move water from the soil into themselves and thence to the air without the wasteful expenditure of energy that this would otherwise entail. Vascular plants (i.e. those with water-carrying vessels) are rather like irrigation channels diverting through themselves water which would, in any case, evaporate from soil to air. How is it done and done so effectively that even a forest giant with leaves many metres from the ground is amply supplied with water? Briefly, the process depends on the difference between the mean energy content of liquid water and water vapour. Liquid soil water has a higher mean energy content than water vapour, so, just as heat moves along a gradient from higher to lower temperatures, water moves along an energy gradient from media with a higher mean energy content to those with a lower mean energy content. But why through a plant?

Figure 18 indicates how the soil/plant/atmosphere system works. It can be seen that because the mean energy content of water in a plant is intermediate between that in soil water and that in atmospheric water vapour, water can diffuse along the gradient from soil to air via the plant. The figures on the diagram indicate the so-called **water potential** of the water and vapour at each stage and are expressed in negative atmospheres. Water potential is the difference between the mean energy content of water at any point in a system and the mean energy content of *pure* water, measured at the same temperature. Now, if a plant is going to be efficient at transporting water across the system it must be adapted especially to any characteristics of resistances (1) and (8) which may vary widely from place to place and from time to time according to the vagaries of climate and soil. Of course, movement of water across this system means that work is being done, and this work is a function of the proportion of incident solar radiation available to do it. The proportion varies widely geographically. In southern England in summer about 40 per cent of the incident solar radiation is available,

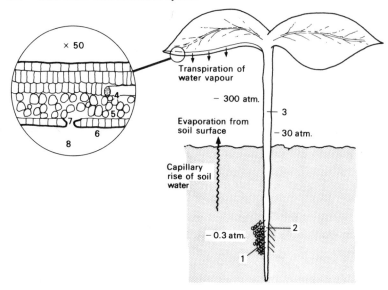

Figure 18. The soil/plant/atmosphere water potential gradient. Note that the figures in negative atmospheres are average examples and would be highly variable from place to place and through time (see text). 1-8 are the resistances which water must overcome if movement is to be maintained along the gradient. (1) soil water; (2) resistance at endodermis; (3) resistance in woody vessels (xylem); (4) resistance to passage into leaf cells; (5) resistance to entering air spaces; (6) resistance to passage through cuticle; (7) resistance to passage through stomata; (8) resistance presented by the atmosphere.

whilst at Manoas in Brazil it is 80-90 per cent throughout the year. At a place like Aswan in Egypt, however, with little or no water to evaporate, most of the solar radiation is used to heat the air, resulting in a hot desert (Figure 19).

Whether water molecules will move rapidly or slowly into the air depends very much on the resistance (8) they encounter. This depends in turn on the water potential of the atmospheric water vapour, which is determined by its vapour pressure. Vapour pressure is a function of the amount of water vapour and how much the vapour molecules are moving (i.e. their kinetic energy). The resistance encountered by water vapour in moving from plant to atmosphere will obviously be less if the vapour pressure is low, so **potential evapo-transpiration** (PE) will be greater in air of lower relative humidity. At 20°C, for example, air of 85 per cent relative humidity has a water potential of approximately —220 atmospheres, whilst at the same temperature air of 50 per cent relative humidity has a water potential of nearly —1000 atmospheres.

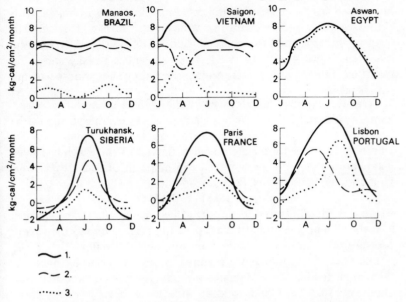

Figure 19. Annual regime of energy-budget components at various localities. (1) solar and infra-red radiation from the earth and atmosphere; (2) heat transferred by evaporation or condensation; (3) heat transferred to or from the surface by condensation or turbulence. (Diagrams from Gates (1962), *Energy exchange in the biosphere,* Fig. 9. Data from Budyko, 1956, *The heat balance of the Earth's surface.* Copyright Harper & Row 1962.)

What happens when relative humidity is 100 per cent and water cannot diffuse along the water potential gradient at all? Plants then may do one of two things if their evolution has fitted them to: either they can exude water as a liquid (a phenomenon known as **guttation**) or they can raise the internal temperature of the leaf above that of the atmosphere by increasing their respiration. Guttation seems to be a fairly widespread facility—tomato, nasturtium and many grasses are vigorous guttators when conditions are right—but it is in the trees of the hot, wet forests that it is particularly important. For example, the coco tree or taro of south India has been recorded as exuding as much as 200ml/day from only one of its metre-long leaves. Drip-tips on these leaves appear to have evolved to shed the copious amounts produced by this and many other species.

In still air, diffusion of water vapour from the leaves increases the vapour content of the immediately adjacent atmosphere, raising its water potential and thus increasing resistance. Wind removes this humidified layer and preserves the gradient. Thus in environments with abundant water and frequent low windspeeds, such as tropical rain forests (where over half the recorded windspeeds at ground level can be less than 5 m.p.h. (Watts 1955)) this problem is usually solved by guttation. In dry windy places, on the other hand, low windspeeds are beneficial.

From the practical point of view, to have some quantitative assessment of this vital habitat factor of potential evapo-transpiration (PE) would be very useful. A number of workers have attempted to derive empirical formulae from meteorological data, and one which has given good correlation with the observed consumption of water by various temperate crops is that of Penman (1948):

$$Emm/day = 1.4 \, (1 + 0.17u) \, (e_s - e_a)$$

where u is the windspeed in m.p.h. at two metres above the surface, e_s is the vapour pressure at the evaporating surface and e_a the vapour pressure at a standard distance from the source of water.

The kind of formula above requires, of course, accurate measurement at particular sites of data not normally given in climatic statistics. Other workers have had to deal with much less accurate, more generalised data in attempting to derive indices of potential evapo-transpiration. It is difficult to say how valid these are. As Watts (1971) points out, 'It has not yet been proved that differences in rates of PE can be used to explain more precise changes in patterns of natural vegetation'.

3.5 The liability of plants to moisture stress

Rainfall figures alone are a poor index to the amount of water supplied to the soil and its availability for growth. The water which is actually absorbed by the soil may be much less than rainfall figures suggest. It is reduced by evaporation from the vegetation and surface debris and from the soil surface before it is absorbed. When the speed of infiltration is low, as in clay soils, the losses can be considerable. This accounts for the fact that in arid regions clay soils are much drier than sandy soils. Similarly, damp soils are slower to absorb water than dry soils. Thus, in well-watered regions clay or humic soils have slow infiltration rates and lose more available water than sandy soils. Once absorbed, water in soil exists in four states, only some of which is available to plants, as shown in Figure 20.

As long ago as 1916, in a classic piece of work, Briggs and Shantz showed that a wide variety of plants could reduce the capillary water of the soil to a point where its resistance to movement became so great that its mass flow into the plant ceased and the leaves wilted. The

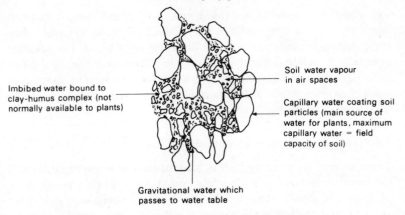

Soil water vapour
in air spaces

Imbibed water bound to
clay-humus complex (not
normally available to plants)

Capillary water coating soil
particles (main source of
water for plants, maximum
capillary water = field
capacity of soil)

Gravitational water which
passes to water table

Figure 20. Distribution of soil water within soil particles ($\times$ 5).

wilting persisted until new water was added to the system. If this
interval was too long the wilting became irreversible and fatal. The
amount of soil water at the point at which wilting was observed was
termed the permanent wilting percentage (PWP), expressed as a
percentage of the dry weight of the soil. Although this might be 10 per
cent for a sandy soil and 20 per cent for a clay soil, significantly in
terms of water potential it could be the same for both, having a mean
value around -15 atm and generally lying between -15 and -20 atm
(Kramer 1949).

Wilting occurs in all plants in natural systems at some time or other,
although it might not be visible by the drooping of stems or leaves.
Physiologically it is defined as occurring whenever there is a water
deficit in the plant resulting from an excess of evaporation over water
absorption. Where the possibility exists that the PWP may be reached
in the soil regularly, plants must have adaptations of morphology,
physiology or life-history to enable them to survive. Whether soil
moisture falls to the PWP or not is a function of soil texture and
chemistry, and of climate. The latter controls the supply of moisture
and the atmospheric water potential which determine the evaporation
rate. Climate also affects water uptake by its effects on soil temperature.
The poleward range of many tropical and sub-tropical species is limited
by the difficulty they experience in absorbing sufficient water at lower
temperatures. Also, it is significant that many plants in the middle
latitudes have strongly developed drought-resistant features even though
water may be abundant during the cold season. Soil texture and
chemistry control the quantity of water absorbed and retained and the
way it is made available for absorption. The chemical composition of

soil water—its **osmotic potential**—may, in some cases, play an important part in determining its availability. This is very true when there are high concentrations of dissolved salts as in the alkaline (solonetz) or saline (solonchak) soils. For plants to maintain a water potential gradient between the soil water and their cells they must lower their own potential by maintaining a stronger concentration of solutes in their cell sap than that which exists in the soil.

3.6 Plant categories in relation to water

There is insufficient space in this account even to begin to outline all the ways in which evolution has fitted plants to avoid the consequences of moisture stress. Most textbooks on plant physiology give good accounts and an excellent introduction will be found in Sutcliffe (1968).

Briefly, plants are conventionally classed into broad categories in relation to water supply. These are:

(a) hydrophytes, adapted to live in open fresh water or permanently waterlogged soils;

(b) hygrophytes, which exercise little or no control over water loss;

(c) mesophytes, which can withstand wilting for short periods;

(d) xerophytes, which show adaptive features of morphology, life history or physiology or all three to withstand dangerous water stress; and

(e) halophytes, adapted to high osmotic potentials in soil water.

Although this is a useful classification it must be admitted that there is no absolute method of distinguishing between the five types. Plants span a broad spectrum in their adaptation to water supply, and many can withstand considerable fluctuations without permanent damage with little more than rudimentary defences against water loss. This is the case with many so-called 'mesophytes' which can adapt by changes in physiology. Then, many 'xerophytes' have transpiration rates as high as or even surpassing those of many 'mesophytes' when water is freely available. There are also unexpected anomalies in the distribution of these types. For example, many of the wet heath plants of Northern Europe are xerophytic in character yet occupy habitats with abundant water supply (Seddon 1974).

3.7 Terrestrial moisture gradients

Three moisture gradients are of importance in the evolution of geographical range in this respect:

(a) gradients towards the centre of continents in mid-latitudes;
(b) gradients from humid sub-tropical regions to the sub-tropical deserts on continental western margins;
(c) gradients from humid tropical regions to the tropical deserts which are gradients both of increasing length of dry season and decreasing rainfall.

In south-east Asia, eastern central America, south-east Brazil and south-east Africa, although the decrease in moisture is not severe, the liability to lowered soil temperatures at certain times of the year appears effectively to limit the range of many rain forest species. This is a good example of the way in which ecological factors may act in concert. Another is represented by gradient (c) above. Experiments in West African forest reserves have shown that many rain forest hygrophyte species are, in fact, quite capable of growing beyond their normal range limit, but only when fire is excluded. They seem capable of withstanding limited drought but are intolerant of fire. The vulnerability to fire is, of course, partly a function of the vulnerability to drought.

3.8 Mineral nutrients

Table 2 sets out the major physiological functions and environmental effects of the most important plant nutrients except water. It may be assumed that where plants are growing successfully there will be few, if any, major deficiencies of mineral nutrients, but the vegetation may well have been selected by the reactions of the plants composing it to the presence or absence of certain minerals, or to their excess or deficiency. However, there are many habitats where mineral nutrient deficiency sets severe limits to growth. Although these habitats may well have plants growing on them, the communities they constitute will be impoverished and well below the optimum production of which they are capable (Proctor 1971).

In spite of the accumulation of knowledge in relation to cultivated plants and their mineral needs since the first classic experiments by Liebig, our knowledge of the mineral needs of natural vegetation is still very poor. As was explained above, there is no natural or man-made ecosystem in which the production of the vegetation approaches the net potential photosynthesis. To what extent the gap in natural ecosystems which are not strictly limited by some other factor is due to inadequate mineral supply—perhaps only a single mineral needed in minute quantities—is not known. Certainly, the fact that agricultural systems which replace natural ones on the same site benefit from fertilisers

Table 2 The major mineral nutrients

Nutrient	Physiological function	Sources	Environmental effects
Macro-nutrients			
Oxygen	Respiratory metabolism	Photosynthesis of green plants	Important determinant of plant distribution in relation to soil aeration. Where soil is badly aerated or permanently waterlogged anaerobic conditions prevent root growth in plants without special adaptations and toxic materials, e.g. H_2S, may be generated.
Carbon dioxide	Source of carbon in photosynthesis	Decay, respiration and the oceans release some 10^{12} tons/year	Some slight natural variations of atmospheric concentration but with little effect on plant growth and distribution. Important ecological effects on soil acidity. Total concentration of CO_2 (320 ppm) sets ultimate limit on photosynthesis. Plants can increase photosynthesis up to $3 \times$ normal with increasing concentration.
Nitrogen	Essential element of proteins. Can only be absorbed in fixed form (NH_4, NO_2, NO_3)	Drawn from the atmosphere by a host of microbes, and lightning	In most well-aerated soils of intermediate acidity fixed N usually freely available. Deficiencies associated with cold, wet soils, very porous soils and tropical soils where vegetation cover is cleared. Where destruction of organic matter is slow or dead material highly lignified, acid 'mor' peat may accumulate as humification (Figure 16) may be inhibited.
Sulphur	Essential for protein synthesis and vitamin synthesis	Sulphates in well-aerated soils, pyrites and gypsum in arid lands, H_2S and reduced sulphur in airless soils	Cycled rapidly by micro-organisms similarly to nitrogen. 'Downhill' losses replaced by weathering, airborne dust, salt spray and volcanic gases. In arid regions strong concentrations of $-SO_4$ ions exist which select for tolerance. Pollutant sources—some 146,000,000 tons annually of SO_2—increasingly added to biosphere.
Phosphorus	Incorporated into many organic molecules, essential for metabolic energy use	Fe, Al, and Ca phosphates; free anions in solution (H_3PO_4 in acid. HPO_4 in alkaline conditions)	Great differences in demand between species and hoarded tenaciously in most ecosystems. Cycled on a world scale with downhill losses replaced similarly to sulphur above. Oceanic reservoir returns deep-water reserve along cold currents via plankton, fish and guano of fish-eating birds (see Figure 16).
Calcium	Essential to metabolism but not incorporated into fabric molecules of living matter	Feldspars, augite, hornblende, limestone, and sulphates and phosphates in arid lands	Strong selective effects in all habitats—lakes, marshes, grasslands, forests, rock outcrops. Important determinant of prime physico-chemical characteristics of soil. Antagonistic to toxic effects of K, Mg and Na. Retention of ions by colloids closely related to climate, especially rainfall.
Potassium	Essential to many metabolic reactions, especially protein-building and trans-phosphorylation	Feldspars, micas, clay minerals	Deficiency has marked effects on carbon assimilation thus lowering production and biomass. Certain crops—beet, cotton, vine, legumes—are very sensitive.

Table 2 *continued*

Nutrient	Physiological function	Sources	Environmental effects
Magnesium	Vital constituent of chlorophyll	Biotite, olivine, hornblende, augite, dolomite, and clays of the montmorillonite group	Excess produces serpentine barrens, e.g. in California, Spain, New Jersey, southern Urals, Japan, New Zealand. Natural climax is replaced on these by impoverished, often scrubby vegetation commonly with specialised ecotypes, e.g. *Quercus durata* in California (see Proctor 1971).
Micro-nutrients Iron	Oxidation and reducing reactions in respiration	Iron silicates, iron sulphates, free ions chelated with organic molecules	Calcareous or alkaline soils may be deficient as iron may be precipitated as insoluble hydroxides. May also be deficient where copper or manganese is present to excess. Vines and fruit crops may be easily affected by iron deficiency.
Manganese	Minute amounts needed for certain enzymatic reactions	Ferro-magnesian minerals. Absorption dependent on other metallic cations	Deficiencies noted in mid-latitudes especially. Tropical soils, especially feralites (see pp. 133-4), may have excess manganese which has toxic effects.
Zinc	Enzymatic metabolism	Zinc-bearing vein minerals	Often leached out of the soil profile in acid soils. May be insoluble in alkaline soils. Certain species, e.g. *Viola calaminaria* of the Harz Mountains in Germany, are endemic to zinc-rich soils.
Copper	Essential for respiratory metabolism	Copper-bearing vein minerals	Deficiency frequent in alkaline soils. Any excess has strong selective effects, e.g. in Katanga, the 'copper flower', *Haumaniastrum robertii*, has 50 × the normal copper content in its leaves; also *Becium homblei* cannot germinate without 50 ppm of copper at least in the soil. The latter is a reliable prospecting index for mineral veins.
Boron	Necessary for successful cell division during growth	Soluble borates are the only assimilable form	May be leached out in acid soils. Some crop plants—beet, potato, cauliflower—show considerable sensitivity to any deficiency.
Molybdenum	Essential for nitrogen fixation and assimilation	Vein minerals	Deficiency in acid soils frequent and also in certain tropical soils on ancient land surfaces.

Note: i) The above list is not complete; there are many other minerals needed in minute quantities which may be involved in the metabolism of most plants or particular groups of plants; ii) Although vein minerals may be important sources of nutrients locally, it is now clear that most non-gaseous nutrients are leached into the oceans continuously by the reaction of sea water with crustal 'hot spots' and thus most continental sedimentary rocks are contaminated with these initially. Most basaltic rocks contain the full suite of nutrients essential for growth but some granitic types may be deficient in certain micro-nutrients (Francis 1976).

suggests that some of the limitation may be due to this. Moreover, the discovery of the phenomenon of 'trace-element deserts' (Anderson and Underwood 1959), where production is even lower than their meagre rainfall suggests it should be, lends support to the idea that other less extreme systems may be similarly limited. Thus in one experiment in Queensland the addition of minute quantities of molybdenum increased grassland plant production dramatically (Stålfelt 1972).

Not only is knowledge of the mineral needs of true nutrients inadequate: so are the effects of those elements like aluminium, chlorine and silicon with which plants have perforce to deal. As these elements affect the soil chemistry and structure and may be absorbed and cycled in significant quantities by plants, they may divert a great deal of energy from photosynthesis which would otherwise be used for growth. (Unlike the passive uptake of water, mineral absorption is essentially an energy-using process—see Chapter 5, p. 82).

Aluminium, for example, is one of the most common soil elements and it has been shown to have marked toxic effects, especially in strongly acid soils where it is easily mobilised. In fact, there now seems to be sufficient evidence to suggest that the categories of 'calcifuge' and 'basiphilous' plants, i.e. plants which avoid or are linked to calcareous soils, contain many species which are not reacting directly to the excess or absence of calcium but to the presence or absence of mobilised aluminium (Clarkson 1966). Species which are tolerant of free aluminium appear to have some mechanism which rejects it, but what this might be is uncertain at present.

It is clear that correlations between plant distributions and mineral supply are not easily established. It is usually only under carefully controlled experiments that any firm conclusions can be suggested about the very complex relationships between plant physiology and mineral distribution. The latter is, in any case, only one aspect of the nature of soil which in itself represents an ecosystem as complex as the one it supports and we shall look at the relationships of the soil/plant system as a whole in Chapter 5.

In conclusion, it can be said that if mineral resources represent the 'fine tuning' of plant distribution patterns, the 'major waveband' is undoubtedly represented by water availability. Of all the materials dealt with in this chapter, it is this which supplies one of the prime keys to the understanding of vegetational geography. Wherever moisture availability is a limiting factor, the biomass of the community which can be supported, the complexity of its layering, the assemblage of its species, the growth forms which they adopt and the percentage of the ground they cover are strictly governed by the availability of this essential nutrient. It is for this reason that the selective pressure exercised by water availability in plant evolution has been so powerful as to produce vegetation of similar appearance and patterns, in locations widely separated geographically and composed of species belonging to genera and even families bearing little or no relation to each other. In the lowlands of the middle and lower latitudes this factor dominates almost all the other ecological components as an explanation of vegetational biogeography.

References and Further Reading

General

*Odum, E. T. and H. T. Odum 1959. *Fundamentals of ecology,* 2nd edn. New York and London: McGraw-Hill.

*Stålfelt, M. G. 1972. *Stålfelt's plant ecology.* London: Longman. (Both the above texts are replete with excellent examples of the principles discussed in this chapter.)

*Watts, D. 1971. *Principles of biogeography.* London and New York: McGraw-Hill.

Water relationships

Durham, R. T. and P. M. Nye 1973. The influence of soil water content on the uptake of ions by roots. *J. Appl. Ecol.* **10,** 585-98.

Kramer, P. J. 1949. *Plant and soil water relationships.* New York and London: McGraw-Hill.

Kramer, P. J. 1962. The role of water in tree growth. In Kozlowski, T. T. (ed.) *Tree growth.* New York: Ronald Press.

Penman, H. L. 1948. Natural evaporation from open water, bare soil and grass. *Proc. Roy. Soc. Arts* **193,** 120-45.

Penman, H. L. 1970. The water cycle. *Sci. Am.* **223(3), 98-110.

Seddon, G. 1974. Xerophytes, xeromorphs and sclerophylls: the history of some concepts in ecology. *Biol. J. Linn. Soc.* **6**(1), 65-87.

**Sutcliffe, J. F. 1968. *Plants and water.* London: Edward Arnold. (A particularly clear exposition beyond the introductory level.)

Watts, I. E. M. 1955. *Equatorial weather.* London: University of London Press.

Nutrients and nutrient cycling

*Anderson, A. J. and E. J. Underwood 1959. Trace element deserts. *Sci. Am.* **213**(1), 97-112.

Chapman, S. B. 1967. Nutrient budgets for a dry heath ecosystem in the south of England. *J. Ecol.* **55,** 677-89.

Clarkson, D. T. 1966. Aluminium tolerance in the genus *Agrostis*. *J. Ecol.* **54**(1), 167-78.

Francis, H. 1976. Report in *New Sci.* **69**(987), 329-32. (A valuable summary of recent discoveries in relation to oceanic 'hot spots'.)

Gale, J. 1972. Availability of CO_2 for photosynthesis at high altitude: theoretical implications. *Ecol.* **53**(3), 494-7.

Hemphill, D. D. 1972. Availability of trace elements to plants with respect to soil-plant interactions. *Ann. New York Acad. Sci.* **199**, 46-61.

Lindsay, W. L. 1972. Influence of soil matrix on the availability of trace elements to plants. *Ann. New York Acad. Sci.* **199**, 37-45.

Ovington, J. D. 1962. Quantitative ecology and the woodland ecosystem concept. In Craggs, J. B. (ed.) *Advances in Ecological Research*. New York: Academic Press.

*Ovington, J. D. 1965. Organic production, turnover and mineral cycling in woodlands. *Biol. Rev.* **40**, 295-336.
(These papers by Ovington contain standard models for the elucidation of nutrient cycles in wooded vegetation.)

Proctor, J. 1971. The plant ecology of serpentine. *J. Ecol.* **59**, 375-410.

*Rodin, L. E. and N. I. Bazilevich 1967. *Production and mineral cycling in terrestrial vegetation*. Edinburgh: Oliver & Boyd. (Standard work on nutrient relationships in the world's major land ecosystems.)

*Rorison, I. H. (ed.) 1969. *Ecological aspects of the mineral nutrition of plants*. Oxford: Blackwell.

Siever, R. 1974. The steady state of the Earth's crust, atmosphere and oceans. *Sci. Am.* **230(6), 72-82. (An excellent account of the integration of the biosphere and recommended as first reading to expand the material of this chapter.)

4

Ecosystem conditions and plant distributions

The incorporation of energy and materials into the bodies of plants composing land ecosystems takes place within a framework of circumstances which are very much conditions of place. We can list the main conditions which influence growth, but are not themselves directly incorporated into the bodies of plants, as heat, soil type, fire, wind and snow cover. Like many of the factors dealt with in the last two sections these conditions of plant existence have a differing impact at the species level from that at the level of the ecosystem as a whole. For example, a forest may spread continuously across large areas of the earth's surface. However, it will be made up of different species from place to place according to the way in which the plants composing the forest respond to the different conditions they encounter. It is at the extremes of the range of an ecosystem type—forest, grassland, desert or whatever—that the conditions listed above may become important at the ecosystem level. This is especially true if, either by themselves or in combination with changing energy or materials availability, they limit the photosynthetic production needed to sustain a particular ecosystem type. This can be illustrated by the altitudinal change of vegetation on high mountains in low and middle latitudes. In these mountainous areas there might be enough light energy and nutrients to sustain tree growth beyond the tree line. However, the lack of heat limits the ability of trees to use the energy and materials to sustain their growth and their place is taken by low shrubs and herbs. In the case of particular species, their distribution is largely *individual* within ecosystems. This is because their genetic material has been selected by evolutionary processes to respond in a particular way to habitat conditions they encounter.

4.1 Heat, temperature and plant life

All biological processes are affected by heat conditions. Heat is a form of vibrational energy stored in matter as the invisible random motions

63

of the atoms and molecules of which it is composed. A measure of this state of agitation of the separate parts (their thermal energy) is its temperature, recorded by a thermometer. Heat flows from the substance —gas, liquid or solid—to the thermometer or vice versa until temperature of thermometer and substance are equal. An actual temperature, for example a degree Celsius ($°C$), is expressed as a fraction of the difference between two fixed and accurately reproducible temperatures, i.e. the freezing and boiling points of water at standard atmospheric pressure in this case. However, physicists may use other more precise scales than this one, based on differently defined points.

Physiological processes in plants and animals are governed in their relation to temperature by the 'Van t'Hoff rule'. This states that a rise of temperature of $10°C$ usually causes a doubling or tripling of the rates of chemical reactions. This rule ceases to apply to life processes, however, once the lethal temperature of protoplasm is reached. At this point (which varies according to species but is usually between $38°C$ and $45°C$), damaging reactions become increasingly important. The thermal energy of the molecules composing living material becomes such that beyond this temperature their structure is increasingly damaged, i.e. they **denature,** and life processes can no longer continue.

The Van t'Hoff rule applies *only* to reactions directly under the control of *internal* temperatures. It does *not* apply to the variation of *external* temperatures. Thus, if the mean annual temperature of a plant habitat were to increase by $10°C$, although the rates of plant metabolism might well increase, it would not produce the doubling or tripling of rates suggested by the Van t'Hoff rule. Plants, like animals, have means of controlling their internal temperatures. For example, in a tree practically all the heat control is exercised by its leaves, especially through convection and transpiration, with radiation being less important. Thus, in vegetation like the tropical wet forests where leaves are exposed to frequent high external temperatures, high incident radiation and low windspeeds, the size, spacing, albedo (reflective quality) and placing of leaves on the stem have been suggested as adaptations to energy exchange requirements dominated by the need for adequate cooling. Even the aerodynamic shape of the crown and the degree of roughness that the forest surface as a whole presents to the wind may represent a compromise between the need for adequate cooling and the avoidance of moisture stress. It has also been shown that the leaves of tropical trees have both a higher optimum temperature for their physiological functioning—photosynthesis especially— and higher lethal temperatures ($45°$-$50°C$) than leaves of temperate trees (Brunig 1971).

Where leaf temperatures are likely to be lower, as in the temperate latitudes, the problems of adaptation to the needs of temperature control are not so acute, and the leafing arrangements—branching, mosaic, whether mono-layered or multi-layered—have been more significantly related to needs other than cooling (Horn 1975). In temperate forests with dormant periods of variable length, the photosynthetic function of leaves is paramount, although they undoubtedly still serve the function of temperature regulation.

Although leaf temperatures are not wholly determined by variations in air temperatures, plant growth rates nevertheless show good correlations with external temperature variations (Gates 1968). Consequently at a macro-scale we may observe apparent correlations between plant distributions and various climatic temperature indices such as isotherms. However, we must be very careful not to draw direct causal inferences from these. It is very tempting from the vantage-point of the small-scale map to infer that species X is 'limited' by, for example, a particular average temperature in July if the iostherm for that temperature and the species distribution coincide. To determine whether the apparent correlation is *more* than a coincidence requires close observation in the field and laboratory of many more facets of a species' physiology and behaviour than the apparent response to temperature presented by map evidence.

4.2 Temperature and tolerance range of plants

It can be observed in the field and laboratory that every organism has a range of tolerance of external temperature in which, during its life span, it can remain indefinitely active. The boundaries of this range are set by the minimum and maximum effective temperatures beyond which the organism ceases its activities and may die. Within their normal geographical range most organisms exhibit adaptations in their life cycle and metabolism which provide protection against these extremes during periods of growth and reproduction. Unfavourable periods of low or high temperatures may be passed as seeds or in a dormant state or by increasing the resistance of the tissues. The effectiveness of cold resistance can be quite remarkable. For example, seeds some 10 000 years old found in permafrost silt have been successfully germinated in the laboratory (Porsild *et al.* 1967).

It is practically impossible to give a single 'optimum' temperature to which an organism is adapted. Rather, a species is adapted to a range of temperature optima for different processes, as is shown in Figure 21. The optimum temperature for germination, for example, may be quite different from that for photosynthesis. Although exposure to cold might seem at first sight to be deleterious to plant growth, it has

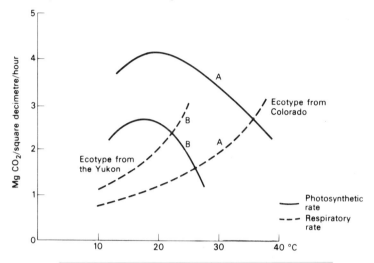

Figure 21. Photosynthetic and respiratory rates in two ecotypes of mountain sorrel (*Oxyria digyna*) (after Mooney and Billings 1961; Copyright Ecological Society of America).

become a vital condition to ensure successful germination in some species and for the initiation of the reproductive cycle in many.

Because all climates experienced by individual plants are local climates which might vary greatly over even quite short distances, plants rarely if ever meet 'optimum' conditions in middle and high latitudes, even for only one metabolic function. Each individual plant works out its own relationship with its habitat, so that the flows of energy between itself and its surroundings are maintained by the controls on its internal chemistry and the individual adaptations of its external form.

4.3 Temperature zonation and geographical range

Four types of temperature zonation have been identified to which the geographical distributions of organisms are critically related (Hutchins 1947). These are:

(a) Control by minimum temperatures where the limits of altitude, latitude and penetration into continents for a large number of species are set by critical minimum temperatures. In this case a parallelism usually exists between the geographical limits of a species and the isotherms for the average temperature of the coldest month or the average monthly minimum temperatures (Figure 22).

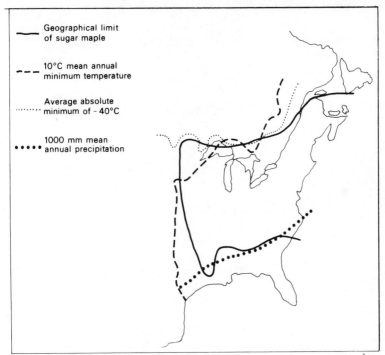

Figure 22. Relationship of the geographical range of sugar maple (*Acer saccharum*) to various climatic indices (after Dansereau).

(b) Control by insufficient heat, i.e. the environment does not provide sufficient heat for the completion of at least one vital activity during the growing season. This applies particularly to the poleward and altitudinal limits of many forest species. The coincidence here would be with the isotherm for the warmest month (Figure 23).

(c) Limitation by excessive heat, which is a condition applying to few species directly, although one of the most important appears to be the vine which seems to be limited by this factor in its extension eastwards from the Mediterranean (Lemée 1967). In most other species excessive heat intervenes to limit geographical area through its effect on transpiration rates.

(d) Limitation due to insufficiency of temporary cold, i.e. where plants need a cold season for successful germination or the initiation of flowering. *Acer saccharum,* the sugar maple, has been cited as being limited by this factor in the southern states of the USA (Figure 22).

Where common thermal boundaries are shared by a large number of species, striking vegetational changes are likely to occur as with tree

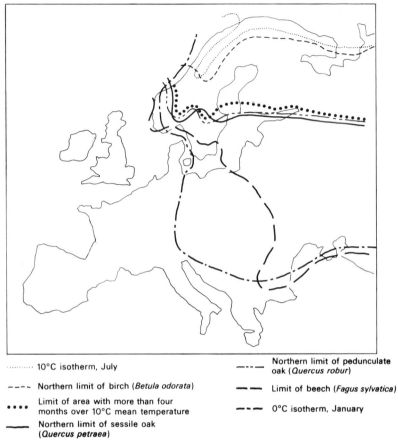

.... 10°C isotherm, July

– – – Northern limit of birch (*Betula odorata*)

•••• Limit of area with more than four months over 10°C mean temperature

──── Northern limit of sessile oak (*Quercus petraea*)

–··– Northern limit of pedunculate oak (*Quercus robur*)

── ── Limit of beech (*Fagus sylvatica*)

–·– 0°C isotherm, January

Figure 23. Relationship between average temperatures and the polewards limits of beech, birch and oak (from various sources).

lines and 'frost lines'. The latter are usually much less sharp than the former and, as is usual with most geographic boundaries, are more likely to be a zone of transition with species being progressively eliminated across it. The striking vegetational changes inland from the Gulf coast of North America and from southern Florida into Georgia and the Carolinas are good examples.

However, in the case of tree lines the fall in temperature with height is not the only factor at work in many mountain areas. In Britain, for example, Pears (1968) has shown that in the Cairngorms the drastic reduction in height which has taken place from the *potential* tree line (600 m—680 m) to the present 500 m or so in recent centuries is due largely to increased wind exposure as the 'wind break' of the great

Caledonian Forest was progressively removed from the sixteenth century onwards. Also the variation in tree lines from mountain to mountain can be seen in certain cases as a result of what has been called the 'Massenerhebung Effect', i.e. the tendency for mountain masses to modify prevailing climate significantly. This can clearly be seen in the Alps, for example, in Figure 24. This is not because the Central Alps are much warmer in summer than the pre-Alps. In fact they are considerably colder at night at the heights reached by trees. Although the effect is not completely understood it is usually attributed to the decreased cloudiness and thus the greater energy available for growth in the Central Alps. The effect can be noted in most mountain masses and is responsible for the highest levels reached by trees at 4500 m in the Tibetan Himalayas.

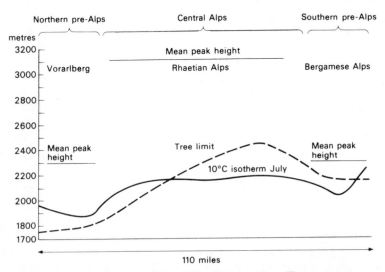

Figure 24. The 'Massenerhebung Effect' in the Alps. (Temperature after Lemée 1967. Mean peak heights and tree limit: author.)

When we examine the polewards tree line we find also that it is rather more complicated than it appears to be on our generalised maps. Correlations are drawn between, for example, a line bounding areas of more than four month above 10°C and the northern limit of birch in Europe. In detail on the ground, however, this polewards 'limit' turns out to be not a line at all but a wide transition zone possibly hundreds of miles deep. In Alaska, there is good evidence that the polewards tree limit is not due directly to lack of summer heat at all but is much more a result of the effect of frost heaving in deeply frozen, poorly drained

soil. Where good drainage exists, the white spruce can be found
growing very successfully hundreds of miles beyond the accepted
temperature limit. Linton (1959) notes the same scattered distribution
of trees in Asia beyond what is normally taken as their limit.

In generally inhospitable thermal climatic regimes, the distributions
of individual species correspond closely with those limited areas within
which their thermal limits of tolerance are not exceeded. In the
northern hemisphere middle and high latitudes, near their cold limits,
plants tend to seek south-facing slopes especially on well-drained soils.
As winter and summer heating regimes vary in the northern hemisphere between slopes of south-east (warmer in summer) and south-west
(warmer in winter) aspect, they may well be located in an even more
intimate relationship to the details of local topography. Such is the case
with the colonies of typically Mediterranean plants found on some of
the limestone escarpments of the Paris Basin, where species particularly sensitive to winter cold cling mainly to slopes of south-western
aspect (see Stott 1970). On the other hand, at their upper limits of
thermal tolerance species are often found on slopes facing away from
the sun or in cooler habitats like those with moist soils or forest cover.
In North Africa, for example, it is possible to find many non-
Mediterranean species in the mountainous areas in precisely such
locations. The communities may even include plants typical of the
peat bogs of Northern Europe (Lemée 1967).

The very general remarks above must not be taken as being true of
all species. In only a few cases are the relationships very clearly
demonstrable. The box tree is one such, where it grows on north-facing
slopes at the southern end of its range in Provence but on south-facing
slopes at the northern end in the Jura. With the majority of plant
species, however, the relationship between thermal conditions and the
limit of geographical range is much less precisely seen. There may
be many factors acting to prevent the further spread of a species than
simply that conditions are 'too cold' or 'too warm'. This is yet another
correlation too easily made, perhaps, when the facts are viewed from
the vantage-point of the small-scale map.

4.4 The flexibility of plant response to thermal variations

The genetics of most plant species allow them to a greater or lesser
extent to vary their morphology and physiology in response not only to
the differing ecological conditions encountered in their range but also
to the way in which conditions may change over time in any one place.
For example, the seasonal change of temperature in latitudes above the

sub-tropics is associated with rapid, intermittent and uneven oscillations of the Polar Front, sudden penetrations to lower latitudes of very cold air or the rapid onset of frost. Thus, the ability to withstand quick chilling and not to be caught 'napping' is one of the most important characteristics essential for survival in these latitudes. It was species that possessed precisely these characteristics which survived the Pleistocene glaciations in Europe and Central Asia, and which still largely account for the vegetation of these regions today.

The rapid resistance to low temperatures which can be developed by some species, particularly evergreens, is striking. In the Alps, for example, at around 2000 m studies of species of heather and of pine (*Pinus cembra*) show that in mid-winter although local minimum temperatures can fall at night to a degree or two below $-20°C$, it would take a fall to at least $-24°C$ to kill the leaves of the heather and a temperature of $-42°C$ to harm the pines. This degree of resistance is developed specifically for local winter conditions. It is not present in summer when a fall to only $-4°C$ would kill heather leaves whilst *Pinus* leaves would die at $-8°C$. However, these temperatures are never experienced in summer, but the plants nevertheless retain a safe 'cold margin' in preparation for the first sudden temperature falls of autumn, when they quickly increase their resistance (Lemée 1967).

The intricate mosaic of habitats created by the play of temperature (and other external conditions) on the varied topography and soils of the Earth's surface may encourage genetic adaptations within species. Either ecotypes may be evolved in geographically isolated populations so that the genetic characteristics promoting survival are favoured at the expense of other characteristics (as seen in Figure 21), or genetic material may be sifted into a genetic gradient. In this last case, at any point along the gradient the species population will possess only those genes for a particular metabolic characteristic most favourable to their survival under the local habitat conditions. It is not easy, except by experiment, to demonstrate that genetic material has, in fact, been sifted out in this way. The classic method is to grow plants from different places side by side under the same conditions to see whether they retain the characteristics of morphology or physiology from their place of origin. A good example is that of Scots pine in Scandinavia. From north to south there is an easily recognisable and very close relationship between the amount of dry matter in the needles of this tree and the growth season length, i.e. number of days with a mean temperature above 6°C. When specimens were grown in the botanical gardens in Stockholm from various points along the geographical range of the plant, each maintained the amount of dry matter in the needles that would be expected from its original location, neatly demonstrating that the gradient characteristics observed represented a true genetic gradient.

4.5 Thermal microclimates and soil conditions

The part that soils play in determining the characteristics of thermal microclimates, although subordinate to slope angle, latitude and other factors, may be of importance in some localities, especially in their effect on seedlings. The first few millimetres above and below the soil surface (where in open sites much of the solar radiation reaches the ground) can provide an environment of extreme severity quite different from the temperature characteristics recorded by a Stevenson screen. In deserts, for example, the temperature in this narrow zone can soar in the hottest times of the year to around 65°C, i.e. well beyond the lethal temperature of protoplasm. In temperate lands, the occurrence of severe ground frosts late in spring, when seedlings are germinating, can produce critical conditions for survival.

On the whole, soils are poor conductors of heat and in most soils daily changes of temperature are detectable only in the first 50 cm or so. Seasonal changes of temperature are detectable, however, through a depth of several metres (Geiger 1957). The kind of heating response that a soil makes to the prevailing meteorological conditions depends on its physical characteristics, especially surface colour (which determines reflectivity), mineral composition, air content and humidity. Soils of good thermal conductivity, such as damp clay soils, heat up less during the day and cool less during the night than dry sandy soils. They are also colder in spring but warmer in autumn. At depth, however, the thermal regimes of the two types are reversed and damp clays are warmer during the day and colder at night.

Where a cover of plants exists, the thermal characteristics near the soil surface are modified in many ways. Vegetation reflects more radiation than bare ground; it traps a layer of air in its branches which absorbs most of the re-radiated energy; and it humidifies the air by transpiration. Consequently, diurnal temperature ranges are much modified when compared to bare ground, and the greatest are found above the surface rather than on it.

Thermal regimes are often strikingly different from one vegetation type to another. If, for example, herbs are growing thickly with leaves parallel to the ground surface, a temperature inversion is maintained between ground and leaves, i.e. the air temperature at the ground is cooler than at the level of the leaves. Where leaves are vertical, this inversion is very near the soil surface. Under closed forests it is at the level of the tree crowns that maximum temperatures are usually recorded, with a secondary maximum at the ground surface if the leaf cover allows sufficient penetration of light. These differences induced by vegetation mean that in describing the thermal microclimate of a wood, for example, standard screen temperatures are of very little use (see Figure 25).

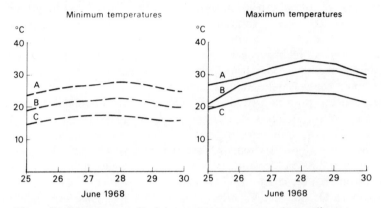

Figure 25. Maximum and minimum temperatures at three adjacent sites during an incursion of continental tropical air, June 1968 (screened thermometers). A—amongst tree branches (height 25 m) in closed canopy oakwood; B—under closed canopy (height 1.5 m); C—thermometer in Stevenson screen 150 m from A and B. Mean wind speed over observation period 5.8 k.p.h. (3.6 m.p.h.). Observation site, Trent Park, Herts, England.

From the remarks made throughout this chapter, it may be concluded that to draw inferences about the relationships between plant distributions and heat energy is quite a difficult task. All such conclusions must be tentative until there is good field and laboratory evidence that they have been fully tested against the way in which plants actually respond to their environments. The fact which should never be forgotten when studying biogeography is that plants and animals are *alive*. They are not inflexible machines destined always to produce the same response to the same circumstances.

4.6(1) Other atmospheric conditions: wind

Although not necessarily a limiting factor in most environments, the movement of the atmosphere is of prime biological importance. The distribution of heat energy, the maintenance of the water cycle and the equalisation of the distribution of gases in the atmosphere involve the wind. However, there are parts of the world where atmospheric movement becomes limiting to plant growth and thus has had an important role as a selective force in the evolution of species and their distribution. We can list briefly the most important effects as follows:

(a) Strong and regular winds on sea coasts and mountains can produce distorted growth forms of various types, e.g 'flag-form' trees, bowed trees, 'krummholz' forms (low twisted bush forms usually of conifers) and elfin forest in the tropics. 'Krummholz' can form even at relatively low altitudes, as the builders of the trans-Pennine motorway found in Yorkshire when coniferous trees planted as snow controls refused to assume a normal habit without themselves being protected from wind. Tropical elfin woodland grows at the upper forest limits and is composed of gnarled, distorted, low trees which are distinctive ecotypes of trees of normal habit found in forest lower down. Extremely windy conditions may prevent growth altogether as in Iceland, many of the Aleutian Islands, the Patagonian west coast, the Falkland Islands and the islands of the Southern Ocean. Significantly, some of these locations, especially in the southern hemisphere, have some of the highest mean surface wind speeds recorded; between 42°S and 50°S, for example, they are above twenty knots in winter.

(b) Strong, regular dry winds can increase the evaporation rate so that in low latitudes especially plants may face extreme moisture stress. A number of workers have shown clearly the significant effects of, for example, the Harmattan wind of West Africa at both macroclimatic and microclimatic levels. In Guinea, at the head-waters of the Milo river, south-facing slopes exposed to the monsoon have over 2000 mm of rainfall per annum and are covered with dense forest, but north-facing slopes exposed to the desiccating Harmattan are savanna-covered and fires are more frequent. In Ghana, the distinctive distribution of bushes and herbs around tree clumps on the Accra plain has been related to the effects of this wind (Lawson and Jenik 1967).

(c) In contrast to (b) above, regular humidified winds can encourage growth, as in northern California where the redwood forests are associated with frequent wind-blown fogs, or they can be limiting to tree growth as in Iceland.

(d) The regular transport of salt spray can be limiting to coastal vegetation. In the tropical forests, for example, there is a frequently-observed relationship at the coastal margins of rain forest of a low scrub composed of salt-tolerant species dividing the forest from the sea (see Figure 40).

(e) Probably the most damaging of all wind effects are those produced by the tropical cyclones. In these violent disturbances where wind speeds may reach 150 m.p.h., many square kilometres of forest may be flattened or badly damaged. Regular hurricane tracks affecting tropical forest may eventually produce forest sections dominated by trees at the pioneer stages of succession, and true forest climax may never be reached. Such areas of 'hurricane', 'storm' or 'cyclone' forest/scrub have been noted in the West Indies, Nigeria, north-east Australia, and Malaya. Some authorities are of the opinion that hurricane incidence has been extremely important in the tropical forest in encouraging the development of new species.

All vegetation considerably modifies wind speed. Over a smooth surface such as a large lake or the sea, the decrease in wind speed is proportional to the logarithm of the height above the surface. Over woodland or even grassland, however, the decrease in wind speed is much greater, owing to the effect of drag over an uneven surface. As noted above, the roughness of the upper surface of tropical forest may be related to the need to increase turbulence and so convective cooling. In general, the rougher the surface, the greater the convection. Thus plants modify wind speeds as part of their adaptation to habitat. Even where only an isolated individual is found, its form is often closely adapted to the effects of the wind.

4.6(2) Other atmospheric conditions: fire

'Wherever plants grow close enough together to carry a conflagration, fire can be a significant component of the biotic environment.' (Daubenmire 1968.)

Even the most unlikely environments can be affected by fire. For example, the wet Tasmanian forests contain eucalyptus species which are normally characteristic of much drier habitats. Their presence has been ascribed to fire, which need not occur more regularly than once in every three or four hundred years to allow them to persist.

If fire is a regular feature of the plant/atmosphere interaction, plants must have special adaptations to render them fire-resistant. Fire-resistance is one of the most important variables in the geographical differentiation of species and in vegetational distribution, especially in the tropical and sub-tropical world. As fires in forests are by far the most damaging and have therefore severely tested the adaptability of trees, fire-resistant features are more conspicuous in trees than in herbs, although the less obvious adaptations of the

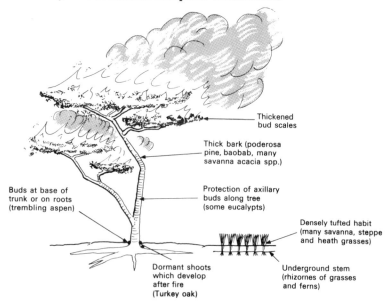

Thickened
bud scales

Thick bark (poderosa
pine, baobab, many
savanna acacia spp.)

Buds at base of
trunk or on roots
(trembling aspen)

Protection of axillary
buds along tree
(some eucalypts)

Densely tufted habit
(many savanna, steppe
and heath grasses)

Dormant shoots
which develop
after fire
(Turkey oak)

Underground stem
(rhizomes of grasses
and ferns)

Figure 26. Some features of plant morphology favouring fire resistance.

latter may belie a considerable capability for surviving fire. Some of the kinds of adaptation to fire are shown in Figure 26.

Most of these adaptations are by fairly straightforward modifications of existing organs, but those that are most closely associated with fire as a selective force are probably connected with the reproductive cycle. Evidence exists that the seeds of some species are positively stimulated by fire, for example those of the valuable African forage grass, *Themeda triandra*. In the USA the seed-cones of some conifers (for example the blackspruce, *Picea mariana*) remain closed until fire-scorched whilst the seeds of competing species are killed.

With Man's activities, fire has become so ubiquitous a condition in many parts of the world that it is almost impossible to define the nature of the true vegetation type. In Australia, for example, the eastern Australian forests contain a full range of resistance from very fire-sensitive to completely fire-tolerant species. These have been sifted by fire (either natural or Man-induced) into protection-dependent and fire-tolerant vegetation-types. North of 32°S where fires are of the surface variety (i.e. they do not burn from crown to crown) the frontier between the two vegetation-types tends to oscillate between eucalypt-dominated, fire-tolerant forests and fire-sheltered raingreen or rain forest types, the border shifting with the impact of fire. In the south-east of the continent, fires are mainly of the very destructive running-crown type with mass-ignition. These are extraordinarily violent with whole stands of trees practically bursting into flame simultaneously. They are so destructive that it is practically impossible to define the true climax vegetation.

The frequency of natural and Man-induced fires in the world's grasslands also introduces an element of doubt into the definition of their 'natural' state. There is certainly good evidence that, given fire protection, in many of them grassy vegetation can be rapidly superseded by trees; for example, by the building of roads as in the Upper Mississippi Basin (Daubenmire 1968) or by the presence of rivers opposed to the direction of the prevailing wind. The last case is well shown in the Llanos of Colombia where the rivers running east-west interrupt fires driven by the prevailing north-easterly winds, and woodland grows on their southern banks (Blydenstein 1967).

The controlled use of fire seems to have been achieved around 500 000 years ago. From that time there is evidence that Palaeolithic man was well aware of its efficacy as a hunting technique. In Spain, for example, at Ambrona around 300 000 years ago, a herd of forty or fifty short-tusked elephants were driven by fire into a swamp and there efficiently butchered. Such use of fire must certainly have had ecological effects. As to how widespread these were it is difficult to say. However, fire-driving as a hunting technique can be traced practically down to the present day (Daubenmire 1968, Sauer 1971).

From 11 000 years ago the first agriculture appears in the record, and it is now clear that it developed in a number of widely scattered places around the world—the Near East, Meso-America, Peru, the Yangtse valley and interior Thailand—within a span of 6000 years or so (Bender 1976). Wherever it developed, fire seems to have been involved as an integral agricultural technique and remains so in many parts of the tropical world to the present day. 'Slash-and-burn', shifting cultivation and 'swidden' farming all employ burning as (a) a quick method of land clearance, and (b) a means of unlocking nutrients from dead plant material. Whether it will actually increase yields depends very much on climate. If conditions are too dry much of the nutrient material in the ash may be blown away. Where rainfall is abundant and reliable, however, the increase of yield can be spectacular (as much as 100 per cent in some experiments in the USA). On the other hand, if burning is done too frequently, the nutrient reserve begins to fall and yields decline. In Africa, the frequency of burning is closely related to population pressure. As population rises, the intervals between burning decrease and with them the ability of the soil to recover after farming.

This point illustrates what is now one of the most important conditions for the operation of ecosystems around the world and the soils which sustain them. There are very few parts of the world now where Man is not one of the dominant influences in the life of the land plants. Before 11 000 years ago Man the hunter-gatherer was limited in

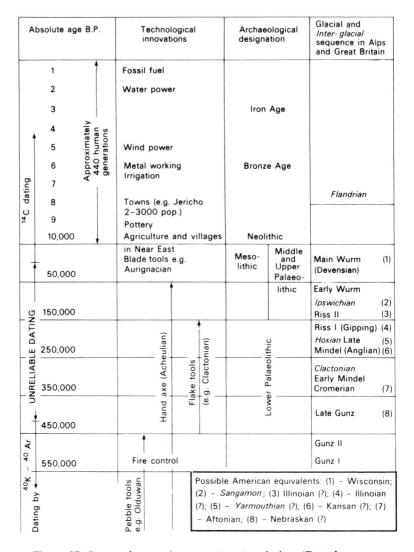

Figure 27. Stages of recent human cultural evolution. (Data from Bender (1976), Goodall (1975), Fleming (1976) and West (1968).)

population by what could be obtained from the wild in the leanest season. At their maximum there were probably no more than 20 or 30 million human beings around the world. With the invention of agriculture, world population has steadily increased to its present 3750-4000 million, and the clearance of land to sustain the human population has by no means finished. In China, the USSR, Brazil and India and on the African continent the process of land colonisation is still rapidly expanding. As to how rapid this process has been, Barbara Bender has provided a telling analogy. If we compare the total span of human life on earth to a period of eighteen days, the hand axe was invented two days ago, fire control yesterday and everything from agriculture to the moon-shots has occurred in the *last half-hour*. Figure 27 sets out the time-scale of these later changes diagrammatically and in Part Two of this book we shall look more closely at some of Man's effects on the vegetation of some of the world's major ecosystem-types.

References and Further Reading

General

Duvigneaud, P. 1974. *La synthese écologique.* Paris: Doin. (Superbly illustrated text by the doyen of continental ecology.)

Geiger, R. 1957. *The climate near the ground.* English language edn, trns. M. N. Stewart *et al.* Cambridge, Mass.: Harvard U. Press.

*Kormondy, E. J. (ed.) 1965. *Readings in ecology.* Englewood Cliffs, N.J.: Prentice-Hall.

Lamée, G. 1967. *Précis de biogéographie.* Paris: Masson. (A mine of data giving access to much continental work.)

**Watts, D. 1971. *Principles of biogeography.* London and New York: McGraw-Hill. (An excellent text copiously referenced.)

Temperature conditions

Brunig, E. F. 1971. On the ecological significance of drought in equatorial wet evergreen forests of Sarawak. In Flenley, J. R. (ed.). *Trans. first Aberdeen-Hull symp. Malesian ecol.,* 66-96. Hull: Dept of geography, Hull University.

**Gates, D. 1968. Towards understanding ecosystems. In Craggs, J. B. (ed.) *Advances in ecological research.* New York: Academic Press.

Horn, H. S. 1975. Forest succession. *Sci. Am.* **232**(5), 90-8.

Hutchins, L. W. 1947. The basis for temperature zonation in geographical distribution. *Ecol. Mono.* **17**, 325-35.

Linton, D. L. 1959. The treelessness of the tundra. *Geography* **48**, 88.

Porsild, A. E., C. R. Harrington and G. A. Mulligan 1967. *Lupinus arcticus* (W) grown from seeds of Pleistocene age. *Science* **158**(3797), 113-14.

Stott, P. A. 1970. The study of chalk grassland in northern France. An historical review. *Biol. J. Linn. Soc.* **2**(3), 173-207.

Wind conditions

Jenik, J. and J. B. Hall 1966. The ecological effects of the harmattan wind in the Djebobo Massif. *J. Ecol.* **54**(3), 167-78.

*King, R. B. 1971. Vegetation destruction in the sub-alpine and alpine zones of the Cairngorm mountains. *Scott. Geog. Mag.* **87**(2), 103-16.

Lawson, G. W. and J. Jenik 1967. Observations of the microclimate and vegetational interrelationships on the Accra plains. *J. Ecol.* **55**(3), 773-85.

*Pears, N. V. 1968. Wind as a factor in mountain ecology; some data from the Cairngorms. *Scott. Geog. Mag.* **83**(2), 118-24.

Fire as an ecological condition

Ahlgren, I. F. and C. E. Ahlgren 1960. Ecological effects of forest fires. *Bot. Rev.* **26, 483-533. (Standard exposition of this subject.)

Blydenstein, J. 1967. Tropical savanna vegetation of the Llanos of Colombia. *Ecol.* **48**, 1-17.

**Daubenmire, R. B. 1968. Ecology of fire in grasslands. In Craggs, J. B. (ed.) *Advances in Ecological Research*. New York: Academic Press. (Standard exposition of this subject.)

West, D. 1965. *Fire in vegetation and its use in pasture management*. Farnham Royal, Bucks.: Comm. Bur. Pasture and Crops.

Human evolution and future prospects

**Bender, B. 1976. *Farming in prehistory: from hunter-gatherer to food producer*. London: John Baker.

Fleming, S. 1976. Man emerging (1): the first 33 million years. *New Sci.* **171(1007), 6-10.

**Goodall, V. (ed.) 1975. *The quest for man*. London: Phaidon.

**Harlan, J. R. 1976. *Crops and man*. Madison, Wis.: American Soc. for Agriculture.

Sauer, C. O. 1971. Plants, animals and man. In Buchanan, R. H., E. Jones and D. McCourt (eds) *Man and his habitat*. London: Routledge, Kegan & Paul.

Scientific American* 1976. Special issue, *Food and agriculture*. **235(3).

(The references particularly recommended provide a valuable summary of the stages by which Man has reached his present state in relation to the biosphere and the prospects which face him in the future.)

5

Soil conditions and their influence on plant distributions

The influence of soils on the kind of plants allowed to grow in a particular place is an extremely complex habitat condition. The soil itself is a system as intricate as the vegetation it supports. Even to outline the numerous geographical, geological, climatic and ecological relationships between the atmosphere, plants and soils would be outside the scope of this book; good accounts will be found in Eyre (1968) and Watts (1971). However, the basic principles of the relationships can be expressed relatively simply in the form of a schematic systems diagram (Figure 28).

5.1 The atmosphere/plant/soil system

Figure 28 shows that the limiting boundary of the system is set by the energy and atmospheric conditions which affect (a) the nature of both plants and soil directly, and (b) the connecting processes between the two contained systems. This figure also indicates the critical importance of the root/soil interface (**rhizosphere**) which forms the direct link between the two systems. It is the response of the root to the soil that determines whether plants become established in an area. Figure 29 shows that the root/soil interface is a much more complex thing than we might think. In the near-root zone, for example, there is always a vast population of micro-organisms positively associated with the higher plant and dependent on the exudation of food from its root. This penumbra of microbes has an important role to play in return for its energy supply. They can exude enzymes (organic catalysts) which break down normally insoluble minerals (e.g. calcium triphosphate) and they protect the plant from attack by other harmful microbes and fungi. The fungi which infest the roots of practically all higher plants—termed **mycorrhizal fungi**—are of two types, endotrophic and ectotrophic. The former feed internally within the higher plant but the latter, which

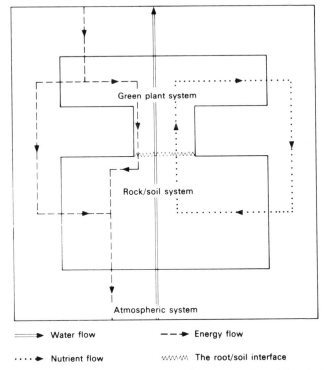

Figure 28. Schematic systems diagram of the atmosphere/plant/soil system.

spread into the soil to feed, are extremely important in maintaining the water and mineral nutrition of the herbs and trees with which they are associated. They form an enormous surface area for absorption, and feed into the higher plant both water and mineral salts.

The successful establishment of a green plant essentially depends on whether and in what form the soil can supply the root with water and mineral salts. As we saw in Chapter 3, the movement of water is largely outside the control of the plant, but that is not so in the case of minerals. Figure 29 indicates that the two flows of water and mineral salts are different. The former process is passive and dependent on the physical state of the soil/plant/atmospheric water potential gradient, but the flow of minerals is an energy-using process; the plant has to do some work for its food.

Although the process is by no means fully understood, it seems to operate as follows. The elements—metals and non-metals—needed for nutrition are absorbed in the form of charged particles or **ions** (metal ions and hydrogen are positively charged—**cations**—and non-metals

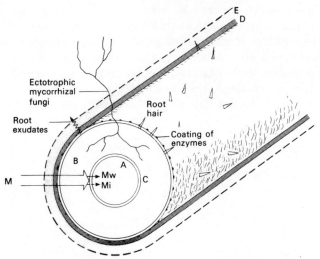

Figure 29. The root/soil interface as represented by the young absorbing zone of the root (× 20). A—stele; B—cortex; C—endodermis; D—near-root zone of electric neutrality; E—zone densely populated by micro-organisms; M—mass flow of water and ions; Mw—mass flow of water; Mi—ion transport by carrier compounds across endodermis.

negatively charged—**anions**). To allow these ions to reach the root in the first place, plants establish, as far as they can, a zone of electrical neutrality near the root so that diffusion can take place across it freely. Ions travel by the same mass flow as water as far as the junction of the stele and cortex of the root—a layer of cells called the endodermis—but here the routes diverge. Water moves across the endodermis by mass flow but the particles become attached to organic compounds within the cells of the endodermis. At this stage there is some limited power of selection of the materials to be carried, but not very much. Charged particles are picked up more or less in the proportions in which they arrive at the endodermis. Consequently, as there are only a limited number of sites available for the arriving ions, a certain amount of 'competition' for sites occurs. Thus an excess of one ion can take up most of the sites and make others relatively difficult to obtain, even if there appears to be sufficient for nutrition from a laboratory analysis of soil. In limestone soils, for example, this can happen with potassium and magnesium where there is an excess of calcium. These processes must obviously exercise a strong effect on plant growth. If from germination onwards a plant has difficulty in establishing its relationship with the ionic content of the soil water or cannot encourage mycorrhizal fungi to infect its roots, it will be at a considerable disadvantage in establishing its seedlings, which may have to compete with the seedlings of many other species. As soils vary widely in their chemical nature the evolutionary process has ensured that this selective force has had an appropriate response in plant genetic material, so that many species are adapted to the establishment of their seedlings in particular kinds of soils with great efficiency (Figure 30).

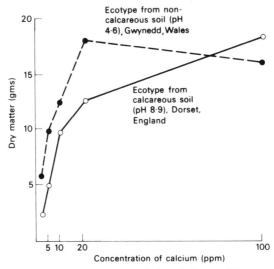

Figure 30. Growth of two ecotypes of sheep's fescue (*Festuca ovina*) in varying concentrations of calcium (after Snaydon and Bradshaw).

Another aspect of the transfer of chemicals across the soil/plant interface is that of oxygen supply. Most well-drained soils are adequately aerated so the roots can obtain oxygen quite easily, but in waterlogged *soil* or very compact soils this is not the case and again natural selection has equipped some species to cope with this situation. The plants of mangrove swamps are a case in point where specialised breathing organs (**pneumatophores**) have developed which project above the mud at low water. Similar organs appear on certain species of the equatorial rain forests. Less extreme examples are the alders and willows of wet soils in Europe. Here oxygen is diffused through the wood from above ground.

5.2 The ionic state of soil water and plant behaviour

The supply of mineral nutrients to plants depends on their presence in solution in soil water. Fortunately for all living things, water, in spite of its ubiquity, is a very unusual substance. It tends to dissolve a great many materials relatively easily, and once dissolved they tend to stay dissolved. This property stems from what is called the **dielectric constant** of water. This is a measure of how well the medium keeps oppositely charged particles apart. The dielectric constant of water is greater than that of almost any other medium. Thus, apart from water vapour, which is always pure, and ice, which is sometimes pure, water in nature is never pure. It is always contaminated by a mixture of positively and negatively charged particles. The measure commonly used to obtain some idea of the ionic state of soil water is that of pH. The pH number is a measure of the concentration of free hydrogen ions and is the logarithm to the base 10 of the degree of dilution. This means that if one gram of a litre of soil water is one hundred thousandths by weight hydrogen ions (i.e. $1/10^5$ or 10^{-5}), its pH is said to be 5. Pure water which has both free hydrogen ions and hydroxyl ions ($-OH$) balanced equally, or neutral solutions, have a pH of 7. That is they contain 10^{-7} grams of $+H$ per litre. Acid solutions have more hydrogen ions and thus numbers less than 7, alkaline solutions have numbers greater than 7. In nature the range can be from 0 to 14 but most soils lie between 3 and 10 and marine environments between 8 and 9.

The pH number can also tell us something indirectly about other elements in the soil. Hydrogen is a very active element. Within the soil water if it comes into contact with a negative charge (or valency) it will hook itself on very quickly. The most likely source of these 'hooks' are the negative valencies not already occupied in the colloidal material (clays and humus) in the soil. In fact, the clay-humus colloids show a 'preference' for attachment of positively charged particles taking them up in the order $H>Ca>Mg>K>NH_4>Na$. Now, if there are a large number of free hydrogen ions in the soil water we can conclude that pretty well all the negative valencies of the clay-humus colloids will be saturated with hydrogen and the other positively charged ions will be free in the soil. This would be the case with a pH of 4 or below. If, on the other hand, the pH number is greater than 8 there will be fewer free hydrogen ions. We can infer one more thing of importance from the pH number. When the number is low, 4.5 or less, and the positive ions are free, they are very liable to be leached out by any water draining down through the soil. It is no accident that soils poor in bases are especially those with low pH numbers. Of course, the pH number cannot tell us anything about the exact amounts of bases available to plants or about their proportions. Other more complex methods must be used to obtain these figures.

What effect, if any, has hydrogen directly? The direct physiological effect is somewhat obscure but it has been observed frequently that significant correlations exist between the distribution of many species and soil pH number. It has also been noted that the effects are most apparent at the seedling stage when plants are establishing their roots. Thus, if a fully-grown mixed ley grass pasture is limed (i.e. made more alkaline) almost nothing happens, but if it is limed *before being seeded* the effects on its composition are dramatic and result in what is to all intents and purposes a monoculture. Grasses not suited to a more alkaline pH as seedlings do not become established.

If we try to find out in the laboratory why some seedlings do better than others in their reactions to the pH of the first few millimetres of the soil surface, a very odd thing happens. Given the same range of pH conditions, species often tend to perform equally well. Figure 31 shows this very clearly in four species taken from acid soils in the Sheffield

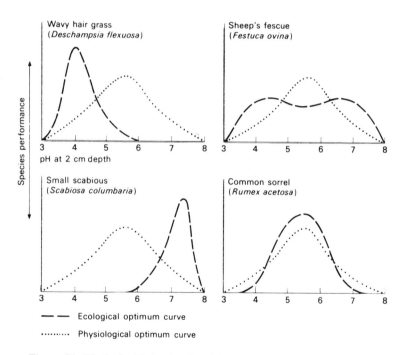

Figure 31. 'Ecological behaviour' and 'physiological behaviour' of four competing species in established pastures in the Sheffield area (from Rorison after Ellenberg). The ecological optimum curves were established from the relative frequency of each species in 340 m² random sample plots in acid pastures. The physiological optimum curves were established for each species from observation of their growth performance under non-competitive conditions in controlled laboratory plots.

area. However, under *competitive* conditions they perform quite differently, as can be seen.

Thus, we must distinguish between 'physiological behaviour' which we can measure in a laboratory, and 'ecological behaviour' which is the way the plant actually behaves in nature. This distinction applies not only to the way plants respond to pH but to many other ecological factors as well. It can be observed that the ecological behaviour in many species, especially near the limits of their geographical range, may be far removed from any physiological optima for them obtained in the laboratory in relation to heat, light, water supply, mineral nutrition and so on.

5.3 The effects of plants on soils

The main effects which plants have directly on soils—as opposed to the indirect effects which they produce by modifying insolation, wind, temperature and other climatic elements—is the way in which, once established, they almost literally 'mould' a home for themselves by altering the soil to suit their needs. This is done in a number of ways but especially by the litter which they drop on to it. This is the energy supply which sustains the organisms of the soil ecosystem. Cut off this energy and the soil rapidly changes its character. It may become compact and airless and liable to erosion. Leaf litter also supplies the vital humus which sustains the clay-humus complex and contains the reserves of soluble nutrients.

The effects of leaf litter on soil character can be quite dramatic. Thus within quite short distances in a mixed oak/beech woodland, the soil type can change from thin, leached acid soils (podzols) under beech to deeper, less acid soils with abundant fauna under oak and holly. An extreme example comes from chalk downland in Sussex which has been planted with pine. Here the soils have progressively gone from shallow, limy chalk soils (rendzinas) to leached, brown forest soils which in some cases have even become podzolised. It can also be demonstrated that when the trees are cut down the process is reversed.

Leaf litter affects the soil chemistry directly by means of the organic compounds it contains. Leaves with abundant organic acids or strongly lignified tissues are not easily eaten by the soil fauna and tend to decompose slowly by wetting, releasing abundant acids and lowering the pH. Leaves with mild (i.e. less acid) organic matter are quickly mineralised and humified and they release abundant bases which are rapidly taken up by the growing plants. There may also be more directly active chemicals released by the vegetation which facilitate territorial dominance by certain species. These were first noted in the walnut tree which releases a compound—juglone—that somehow effectively prevents the growth of competitors. Similar compounds have been demonstrated in many land plants from the lichens to many forest species. These antagonistic reactions between species, and between individuals of the same species, may be very widespread, as we have undoubtedly not identified all of them (example in Figure 32).

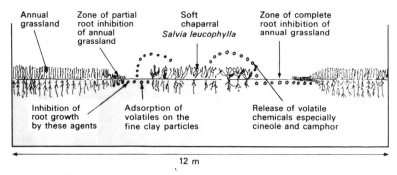

Figure 32. Allelopathic interaction between vascular plants, California (data from Muller).

In a few species the opposite interaction has been proved. For example, the eastern white pine in a New Jersey forest has been shown to exhibit root grafting between adjacent trees. Radioactive tracers injected into one tree turn up in adjacent ones without entering the soil, indicating that the trees are co-operatively sharing resources (Billings 1964).

Plants also prepare the way for each other by building energy and nutrient resources into the soil as a plant **succession** develops. The first colonisers on soils derived from low-nutrient parent material or on very acid or very alkaline soils tend to be low producers. As these draw up minerals from the lower soil levels and involve them in the nutrient cycles, they build up the stock of nutrients in the soil so that colonisation becomes possible for plants with higher productivity.

From all that has been said so far on plant/soil inter-relationships the reader will be aware that to establish causal relationships between plant distribution and substrate chemistry is not a simple matter. Professor Clapham puts it well when he says that 'ecologists find themselves faced by the full and daunting complexity of the plant/soil system when they set out to understand what part mineral nutrition plays in determining the facts of species distribution'.

5.4 Soil classification

To render the task of establishing causal relations between plant distributions and mineral nutrition less 'daunting', it would obviously be of great assistance to have a coherent view of the ways in which soil characteristics change from place to place. However, this is not possible because there are a number of opinions as to what are the most significant characteristics of soils for the purpose of classification. Broadly, these fall into two groups: (a) those which place emphasis on objective criteria based on intrinsic soil characteristics and aim at a zonal application on a large scale; (b) those which treat the soil more empirically and classify it according to its nature as a medium for plant growth. The first group can be exemplified by classifications such as the US Dept of Agriculture's 'Seventh Approximation' and the FAO Unesco system. These rely particularly on the **profile** the soil presents vertically. From the surface down to the unweathered rock, a particular soil will contain a number of more or less distinct horizontal divisions (**horizons**) whose depth, colour, chemical characteristics, etc. can be used to define the **soil type.** As developed in the USSR particularly, the major zonal soil type is presumed to be largely a result of climatic conditions within a major regional climatic type. Many of the terms derived from Russian practice are very familiar to western earth scientists, for example podzol (Figure 51), chernozem (Figure 54) and rendzina. Terms derived from the Seventh Approximation and the FAO/Unesco classification may be less familiar and there is a considerable overlap between them so that a number of terms may describe the same soil type. Thus a grey podzolic soil (i.e. a thin, acid, leached soil with well-marked horizons) is termed an **acrisol** in the FAO system and a **spodosol** in the Seventh Approximation. (See Burnham (1975) for equivalence of terms.)

A number of workers have argued that as holocoenosis is inherent in the soil/plant/atmosphere system such zonal classifications are not particularly helpful in making sense of plant distributions, as in some cases at least it is 'vegetation which determines the finished product' (Eyre 1968). Whether the argument can be carried quite so far as to render zonal classification redundant is doubtful. As Burnham (1975) points out, in the tropics there is still no real substitute for the zonal view if a coherent account is to be rendered of the soils of a large region.

Workers who have tried to present a more empirical approach include Milne (1935), Wilde (1958) and Webb (1968). The views of Milne have had considerable influence in the study of tropical soils especially. Here the most significant factor to be established in relation to a particular soil is the position it occupies (**site**) in the suite of soil types (**catena**) which can be so frequently observed on slopes in the tropics (Figure 33). Within such a catena, loss of nutrients by downslope leaching would be observed at the top of the slope and enrichment at the bottom, with corresponding vegetation changes.

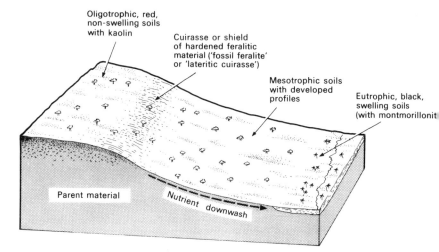

Figure 33. A tropical soil catena. (Terminology after Milne 1935 and Webb 1968.)

The approach of Wilde, on the other hand, is based firmly on the nutrient status of the parent material in relation to forest soils. Thus rocks like sandstone and siliceous shales yield few nutrients when compared with parent material like granitic rocks which may be low in only a few nutrients (phosphorus, calcium and magnesium) or basaltic types which are nutrient-rich. However, in places like Britain or large areas of North America, where the surface is often mantled by glacial drift derived from a hotch-potch of rocks, the classification tells us little about nutrient status.

Webb has attempted a compromise between these two views and established categories similar to those long used by ecologists for aquatic habitats. Thus he identifies eutrophic types enriched by downslope movement of nutrients or on nutrient-rich substrates, oligotrophic types on nutrient-poor parent material or upper slope sites, and mesotrophic or eutrophic-mesotrophic types between the other two. By applying these categories in conjunction with vegetation analysis, Webb was able to achieve a unified classification of the soils and vegetation of the whole of the eastern Australian rain forests.

This study was also able to demonstrate clearly the interaction of the soil/plant system with one of the most important components of the atmospheric limiting system in that part of the world, the incidence of fire. For instance, where the forests are particularly liable to fire, a mature vegetation seems never able to develop, but the particular stage towards maturity reached in any particular place is closely dependent on whichever of the four soil types is present. Those nearest maturity are on eutrophic soils, those furthest from it on oligotrophic types.

It will be clear from this section that to draw simplistic causative inferences about the connections between vegetation and soil distribution is not possible. Only by careful analysis of the ways in which the atmospheric, plant and soil systems interact and are unified at particular places can the proper connections be established.

5.5 Plants and soil salinity

Probably the most extreme soil chemistry conditions experienced by plants are those where saline soils occur. Although there are no completely satisfactory criteria for defining the degree of salinity of a soil, it can be said broadly that where sodium is more than 50 per cent of the total cations in solution the soils are usually strongly alkaline (pH 9 or more), compact and airless, and sodium carbonate forms easily (**solnetz** soils). Where percentage of sodium cations is less, the soils are less alkaline (pH 8.5) and have a good crumb structure (**solonchak** soils). Under arid conditions salt, gypsum or lime crusts may occur on the former types.

According to their response to salinity, plants are conventionally classed into intolerant **glycophytes, glyco-halophytes** able to withstand moderate salinity—cotton, date and barley are examples—and **halophytes** which find their optimum growth condition in saline soils. The common seashore plants such as glasswort (*Salicornia herbacea*) and sea blite (*Suaeda maritima*) are true halophytes.

Physiologically, salinity acts on plants in two main ways. In the first place, the large amounts of dissolved salts lower the water potential of the ground water, making it more difficult to obtain. In addition, the dissolved salts adversely affect nutrient absorption, or even have toxic effects where there are high concentrations of $-Cl$, $-HCO_3$ and $+Na$ ions.

On the whole, ions are absorbed by halophytes in the proportions in which they exist in soil water, but most halophytes have some regulatory controls on salt intake, so that water can continue to move along the water potential gradient. Many, such as rice grass (*Spartina* spp.) and some mangrove plants, can excrete salts via epidermal glands in the leaves; others, like glasswort, regulate the internal concentrations of salts in the cells by increasing their succulence. Many glyco-halophytes, like the shad scale (*Atriplex confertifolia*) and the sage brush (*Artemisia tridentata*), directly curtail the absorption of salts by the roots, although how they do this is obscure.

5.6 Mineral cycles and erosion rates

The cycling of nutrients under the control of plants must obviously have some effect on erosion rates, but until recently very little was known about it. However, from a series of careful studies begun in 1963 on small watersheds in New Hampshire, some accurate figures are now available (Borman *et al.* 1970). Calculations indicate a weathering rate of around $800 kg/ha^{-1}/an^{-1}$ of the granite and glacial till which compose the parent rock. Under the present climatic conditions this would lower the surface by about 0.03 cm/year or 50 cm in the total time since the end of the last glacial epoch (14 000 years in this area). The same studies have also shown how important the nutrient rain from the atmosphere is to the plants. The average figure in total is around $51 kg/ha^{-1}/an^{-1}$, composed of varying amounts of calcium, sodium, magnesium, potassium, chloride, silica, sulphate, nitrate and ammonia. The last three components represent about 75 per cent of the total, and with potassium are clung to tenaciously by the ecosystem.

Obviously, these figures cannot be applied generally to humid temperate ecosystems, but we might speculate from them for other areas of the world, taking into account the rocks and climatic circumstances. In Britain, for example, in the highland zones one might say that weathering rates would be about the same or somewhat lower, but, with the lack of trees, the possibility of trapping the nutrients would be less. In lowland Britain with less resistant rocks yet with relatively frequent and abundant rainfall it might be suggested that the weathering rate could be somewhat greater than on the Hubbard Brook watersheds. Since the removal of most of the forests the erosion rates have probably increased considerably, although there are few visible signs except in certain places where wind removes very light, sandy material.

Gross substrate movement is a normal feature of a number of habitats and there are many species of plants which have evolved adaptations to cope with these circumstances. We may list the main kinds of disturbance as follows:

(a) Movement of unstable, immature soils as in screes and rock-slides which encourages colonisation by pioneer species adapted to forming large, tough mats bound by extensive, deep-seated roots (see Figure 36 below).

(b) Frost-heaving of the surface which may disturb roots and possibly prevent the growth of plants altogether. In moderate conditions although mosses and lichens are intolerant, plants like Arctic willow (*Salix herbacea*) and mountain sorrel (*Oxyria digyna*) can become established and fix the surface against heaving.

(c) In arid lands wind deflation and abrasion damage above-ground organs and expose roots, and wind deposition may bury plants. However, the latter condition may also encourage some species, as with marram grass and the drinn.

(d) Sheet and gully erosion is damaging to permanent vegetation in arid and semi-arid lands but is rare where an established ecosystem exists, even if the ground is bare. With agricultural clearance the rates of erosion can be high. For example, in the USA figures for total erosion under cotton cultivation show that a slope of only 8-10 per cent can produce an erosion rate as high as one centimetre per year.

In less extreme circumstances than those outlined above, the control on physical movement of the atmosphere and the subsoil is a function of the plants themselves. From the pioneer colonisers on any site to the mature vegetation the tendency is for the stability of the surface to increase. Indeed, without this inbuilt tendency to stability, mature communities would not be possible at all. Thus, clearance for agriculture always brings with it the threat of soil erosion whatever the environment. Only if proper conservation techniques are built in to the agricultural system can it be minimised.

References and Further Reading

General

Billings, W. D. 1964. *Plants and the ecosystem.* Belmont, California: Wadsworth; London: Macmillan.

Buringh, P. 1970. *An introduction to the study of soils in the tropical and subtropical regions,* 2nd edn. Wageningen: Cen. Agric. Publ. & Doc.

*Eyre, S. R. 1968. *Vegetation and soils.* London: Edward Arnold.

Hudson, N. 1971. *Soil conservation.* London: Batsford. (Clearly documented account of the practical aspects of soil conservation.)

**Watts, D. 1971. *Principles of biogeography*. London and New York: McGraw-Hill.

Wilde, S. A. 1958. *Forest soils*. New York: Ronald Press.

Soil and plant nutrients

Borman, F. H. and C. E. Likens 1970. The nutrient cycles of an ecosystem. *Sci. Am.* **228(10), 85-92. (An excellent summary of a fundamental piece of research.)

Deevey, E. S. 1970. Mineral cycles. *Sci. Am.* **223**(3), 148-60.

*Douglas, I. 1969. The efficiency of humid tropical denudation systems. *Trans Inst. Brit. Geogs* **46**, 1-17. (Clear exposition of the relationships between erosion rates and ecosystem dynamics in the humid tropics.)

Gibbs, R. J. 1967. Amazon river: environmental factors that control its dissolved and suspended load. *Science* **156**, 1734-6.

Soil classification

Burnham, C. P. 1975. The forest environment: soils. In Whitmore, T. C. *Tropical rain forests of the Far East*. London: Oxford U. Press.

Milne, G. 1935. Some suggested units of classification and mapping particularly for East African soils. *Soil Research* **4**, 183-98.

*Webb, L. J. 1968. Environmental relationships of the structural types of Australian rain forest vegetation. *Ecology* **49**(2), 296-311.

*Webster, R. 1968. Soil classification in the United States: a short review of the seventh approximation. *Geog. J.* **134**, 394-6.

*Young, A. 1974. Some aspects of tropical soils. *Geography* **59**(3), 233-9. (Excellent brief introduction to a complex subject.)

6

Plants in communities and their distributions

In nature, each individual organism and the species population of which it is part must constantly make adjustments to the presence and needs of the other plants and animals which share the land with it. The pressures of living in a community, together with the adaptations to the non-living habitat factors dealt with above, are selective determinants in the evolution of the general range and size of a species. Within this general range each individual plant must establish its own relationships with its neighbours as best it may within the limits of its genes.

6.1 The distribution of individuals in a population

The main ways in which individuals may be distributed in a population are shown in Figure 34. To determine which pattern is present in a population usually involves sampling and mathematical testing. A standard method is to map the distribution of species, preferably in a large number of samples, and match this against a theoretical model of randomness established for the same number of species and samples. The deviation from randomness indicates the kind of pattern present. Another test is to measure the distance between individuals in some standardised way and plot its square root against the frequency, to obtain what is called a **frequency polygon.** The shape of the polygon—symmetrical for random, skewed to the right for uniform, to the left for clumped—indicates the type to which the distribution pattern may belong (Dice 1952).

Whatever the method used, the most frequently observed arrangement, in the vegetation of the middle latitudes at least, is the clumping of individuals. In the tropical rain forests, however, many observers have noted that individuals of the same species tend to be scattered and not clumped together, where forests are composed of the populations of many species. It has not been established whether the scattering is random or the spacing between individuals even.

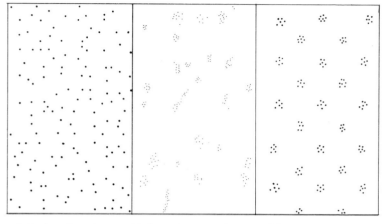

A. Random dispersion   B. Clumped or contagious   C. Regular dispersion
                              distribution   of clumps

Figure 34. Diagrammatic representation of three ways in which
individuals of a species population may be distributed horizontally
within a community. Note that in (A) the distribution has been
established mathematically as random but in certain parts the random
pattern can simulate other distribution patterns.

Environmental factors encouraging clumping include localised water
supply or deficit, slopes, exposure to insolation and the wind, and the
uneven pattern of the chemistry and texture of the soil. It is also a fairly
well-established principle with temperate plants that the distribution
pattern is much influenced by the nature of seeds: heavy seeds tend to
stay together, light seeds to blow away. The last principle is not
applicable to many tropical plants, however, which often have heavy
seeds yet scattered distributions. Undoubtedly the key—or keys—to the
geographical pattern of species populations in complex tropical forests
has yet to be found (Whitmore 1975).

The advantages to those species which are adapted to promoting
close geographical proximity of individuals are considerable. They
include modification of the soil through leaf litter, elimination of
competition by occupation of the available territory, and encouragement of an adequate population of pollinating animals. These promote
the survival capacity and competitive strength of the group and any
disadvantages which may result from intra-specific competition for
nutrients, root space, light, etc. are outweighed. Thus in a beech
woodland in Britain, the leaf litter from many individuals can
accumulate to produce the shallow, acidic soils preferred by this plant
much more effectively than from a single individual. Also the heavy

shade of a closed beechwood canopy effectively reduces competition from other forest trees.

As to how many individuals of the same species may be aggregated together, there are limits which depend on a balance between the growth rate and the ability of the habitat to support it. In each species there is usually an optimum population size at which the advantages of group living are achieved without the disadvantages, and this number is a function of the genetics of the species. If the habitat is not supplying the resources for growth to reach an *optimum* (not maximum) number of individuals, underpopulation may occur and the species becomes vulnerable to competition from those better adapted to the conditions. If the habitat sustains too rapid a growth rate, over-population may occur and competition start in earnest between individuals which may also become liable to disease.

The concept that an optimum population of a species in any particular habitat depends on the balance between the growth rate and the ability of the habitat to support it is known as Allee's Principle. Although this was originally established for animal populations, it applies equally to plants.

The less frequently observed case of a regular distribution of individuals—characteristic of desert shrubs, for example—is usually a result of either intra-specific competition or of positive antagonism between members of the same species. In the first case, as the density of population increases, intensified competition for light, water, minerals, etc. may result in the elimination of many individuals to produce eventually a steady-state condition where the survival rate is equal to the mortality and an even pattern is established. In the second case, positive antagonisms between individuals of the same species have often been demonstrated in the territorial arrangements of animals, but in plants they are much more difficult to prove although they undoubtedly exist. There are, however, some indications that in higher plants interactions favouring the elimination of competition and the sharing of resources territorially between members of the same species may be of considerable importance. A significant example illustrates this point:

One of the most puzzling aspects of plant distributions in the tropical rain forests is why plants of the same species may be widely scattered, and rarely associated together. No completely satisfactory theory has yet been produced (Richards 1973). However, in one non-gregarious species at least, *Grevillia robusta,* there is good evidence that it is the result of a positive antagonism between the parent tree and its offspring. The parent releases into the soil a substance which is water-soluble and which is taken up by the soil micro-organisms. Once armed with this, they effectively prevent the growth of new seedlings. The nature and precise *modus operandi* of this substance are obscure, but its effects have been amply demonstrated in field and laboratory and it undoubtedly frees the adult tree from competition with its progeny (Webb *et al.* 1967). Interactions of this kind may be widespread, but the complex biochemistry involved makes them difficult to describe precisely. Thus, it has been noted that under an established canopy of oak in Britain, acorn germination is fairly rare (Mellanby 1968), but whether there is a mechanism similar to the one above is not known.

6.2 Interactions between species and distribution patterns

At the most elementary levels of biology teaching the close association which exists between certain species is used as a demonstration of the mutual dependence of living creatures. The examples are usually classed under 'parasitic', 'symbiotic' or slightly looser 'commensal' relationships. However, in recent years biologists have come to realise that the symbiotic relationships, i.e. close associations for the mutual benefit of species, are not so much isolated and special cases but fundamental to the very nature of living material.

At the level of whole animals and plants we have already seen (pp. 81-2) the vital role of micro-organisms symbiotically associated with the roots of higher plants, and these may certainly have geographical effects. For example, experiments in the Northern Great Plains of Canada indicate that the unsuitability of prairie soils for pine mycorrhizal fungi is a severe limitation on the spread of trees and may give added sharpness to the forest/prairie transition (Rowe 1967). Similarly, the preference of mycorrhizae of heather, rhododendron and *Vaccinium* (bilberry, cloudberry, etc.) species for soils with a low pH encourages the confinement of these plants to such media. The mycorrhizal fungi are, of course, only a fraction of the enormous population of micro-organisms positively associated with the roots of higher plants. Mycolytic (i.e. fungi-destroying) bacteria and bacteriacidal fungi insulate the plant from infection and their absence sometimes makes difficult the transference of plants from one region to another. This was the case with early attempts to grow wheat and flax in the acid, podzolic soils of the Northern USSR. Generally where pH is strongly acid, bacteria are inhibited and the plants are liable to fungal infection, thus producing a strong selective barrier to colonisation of such media by plants with poor resistance. It can even happen that certain plants may provide a medium which ultimately destroys their own beneficial symbionts and leaves their roots open to attack. This can certainly happen with the European beech whose leaf litter is strongly acid. Conversely in soils with an alkaline pH, fungi find difficulty and a tree may be open to bacterial attack.

Where species are associated in commensal relationships their distribution patterns may not be so markedly related as with true symbionts. Thus in many wet forests, epiphytes which use trees for support may not be associated with any particular tree but generally with trees whose roughness of bark allows them to become firmly attached. In the south-eastern USA, for example, the so-called 'Spanish Moss' lichen (itself the result of symbiosis between a fungus and an alga) festoons the rough oak species but not the smoother pines. (Being without roots in the ground, epiphytic species are very sensitive to humidity conditions and their territorial pattern closely corresponds to variations of microclimate.)

In practice, it is often very difficult to distinguish between symbiotic, parasitic and commensal relationships; and, in terms of distribution patterns, parasitic relationships, although different in their effects, are nevertheless very similar to symbiotic associations. Thus, the strangling fig (*Ficus* spp.), common in the humid tropics, behaves as a harmless epiphyte in the upper branches of forest trees as a juvenile, but when its roots reach the ground it lives up to its name by killing the host with root competition and shade and replaces it in the upper canopy until both parasite and host collapse when the host trunk can no longer provide support.

6.3 Competition for territorial resources

The principle of Gause, established by experiment in 1934, is still the prime rule of species' competition; it states that if two species are competing for the same resources in the same vertical and horizontal space at the same time, one of them must be extinguished. If, on the other hand, they differ in their requirements of space, time or resources (even if some of their requirements overlap) they may co-exist as balanced populations. The major implication stemming from this principle is that any stable community must be made up of interacting species populations *differentiated* as to their need for resources, i.e. with each population occupying a different 'position' or **niche** in the community. The differentiation of niche is a powerful evolutionary factor as the selective advantage to competing populations of an adequate supply of habitat resources far outweighs any advantage which might be gained by direct competition.

One of the most obvious results of niche differentiation in spatial terms is the sharing of vertical living room in vegetation. There are few communities in which some stratification cannot be detected above and below ground. Each species in a community has an optimum position along the vertical gradient of sunlight from the canopy to the forest floor, for example. Beneath the ground differentiation of niche by stratification of root space is similarly apparent. Thus, the life-form of the plant is an expression of this niche differentiation. From the tall canopy trees with their massive and long-lived branches through the smaller shrubs and bushes to the herbs and ground vegetation, the life-form differentiation represents an evolutionary pattern of forms suitable to the gradient of productivity. The trees with their photosynthetic organs in broad sunlight are able to produce sufficient food to fulfil the energy requirements for the formation of woody tissues, whilst, at the other extreme, the herbs with their strictly limited possibilities of production expend little of their precious food reserves on organs of support.

The ways in which horizontal space is shared out amongst competing species are both more varied and slightly less obvious than in the case of vertical space. However, research has demonstrated that in the establishment of plant communities species populations usually become differentiated spatially according to their response to habitat variations. Thus, although their distributions may overlap, the *peak* population of each species is located at a different position in relation to a gradient of change in the habitat from place to place. We saw an illustration of this above in the case of four populations of herbs responding to variations of soil pH. Although each species in the laboratory behaves in the same way in relation to pH, in the field the maximum numbers of each are found at different points on the pH gradient and thus at different places.

The techniques of **gradient analysis** which have developed over the last few decades have been applied very successfully to many communities, especially in the USA (see Whittaker 1970), and have provided a very useful tool in our understanding of the way in which horizontal living space is shared out. Gradients may be of many kinds— climatic, soil conditions, pH, salinity, flooding incidence, slope and so on—all of which act to differentiate one species population from another, each finding a particular niche along the gradient.

There is some debate, however, as to whether the process of niche differentiation as outlined above accounts for species diversity in all communities. Some authorities (e.g. Federov 1966) have argued that in the tropical rain forests niche differentiation has little role in producing community diversity amongst trees, although in other life forms it is probably important. When it is remembered that in the richest of these forests in South-East Asia over 100 tree species per hectare is usual (excluding seedlings), often on practically homogeneous surfaces, it is admittedly hard to see how niche differentiation alone could produce such variety.

The reasons for species diversity in these benign environments seem to be very complex and there is no complete theory which accounts for it. (The reader is referred to Longman and Jenik (1974) pp. 68-74 and Whitmore (1975) pp. 12-40, for a fuller examination of the topic.)

To those familiar with the forests and upland plateaux of Northern Europe, these processes of differentiation to produce complex vegetation are less familiar perhaps than the observation of species dominance. Here the vegetation pattern can be described in terms of 'an oak forest', 'an ash forest', 'a heather moor' or 'a bracken-covered heath'. There are probably two main reasons for this. In the first place, where the habitat is severe—deserts, high mountains, tundra, boreal forest—the pre-emption of niche space by the most successful species is very marked, so that by far the greatest proportion of production within the community is performed by only one or two species. Secondly, in forests such as those in the British Isles, dominance is expressed by one or two species because the Quaternary glaciations removed most of the potential competitors from the scene and the niche space of the casualties has been pre-empted by the survivors, possibly with some further evolution to consolidate the territory thus gained. Where more trees survived, as in the southern Appalachians, dominance is much less marked and territory is shared out according to the rules outlined above.

6.4 The areal variation of species diversity in communities

On a global scale, the diversity of species incorporated into communities is a function of latitude. From the few species of the Arctic tundra, through the temperate forests to the bewildering diversity of the sub-tropical and tropical forests where a single hectare may alone carry more than 100 tree species, the increasing diversity of species with decreasing latitude is one of the fundamental observations of biogeography. At any particular latitude, species diversity amongst particular vegetation types varies from habitat to habitat. Thus the flora in forests with intermediate moisture conditions is usually richer than in very dry or very wet habitats and usually richer than in grasslands. Also forests of intermediate habitats with open canopies are usually richer than forests with closed canopies.

To gain some idea of the degree of diversity in a community, as well as aspects of dominance, niche diversity and competition, a curve of species importance may be drawn. The curve can be established on a variety of criteria—production, biomass, population density, etc.—and plotted as shown in Figure 35. The two curves represent the two main types of distribution found in plant communities. Curve A is characteristic of communities of extreme habitats where the number of species is small. Here we have strongly marked species dominance. Communities which are rich in species usually resemble Curve B with a descending scale of importance so that each species occupies only slightly more or less of the available niche space than the species immediately adjacent in the rank order. Once it is established to which of the types the species diversity belongs, we might then look at the way in which this diversity varies from place to place.

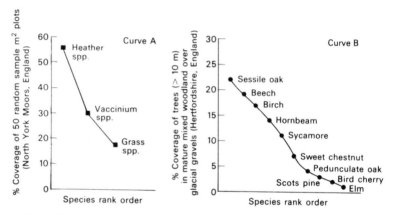

Figure 35. Two types of species importance curve. Curve A is characteristic of plant communities in more extreme environments whilst curve B is more characteristic of complex vegetation. The woodland from which curve B is derived is shown in Figure 50**B**.

Two types of diversity have been distinguished (Whittaker 1970), the first being diversity within communities, which has been called alpha diversity. The second is the degree of change in species composition through space, which is termed beta diversity. Figure 36 illustrates the difference between these two kinds of diversity.

In this example, the most severe habitat, the old pit heap, has a low alpha diversity and this seems to be a general rule in nature. Severity of conditions encourages the natural selection of specialised adaptations and gives those species which develop them a great advantage in the pre-emption of niche space. In this case, the creeping bent grass (*Agrostis stolonifera*), for example, by its densely-tufted habit has good control of water loss and with its strong horizontal shoot can stabilise the surface.

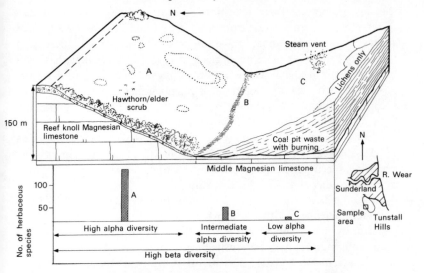

Figure 36. An illustration of alpha and beta diversity. A—limestone pasture dominated by blue moor grass (*Sesleria caerulia*) with many co-dominants and associated grasses, sedges and other herbs. Local co-dominance with other species indicated by dotted areas; B—mixture of species from A with red fescue (*Festuca rubra*) and spreading meadow grass (*Poa subcaerulea*) important; C—skeletal soils dominated by coarse grasses especially cock's foot (*Dactylis glomerata*) and creeping bent (*Agrostis stolonifera*). Few insect-pollinated plants. Steam vents associated with moss (*Polytrichum*). (Fieldwork: author and D. Hall.)

6.5 The development of communities on new sites

The replacement of one plant community by another in an orderly progression—or **sere**—is a readily observable phenomenon in nature. As each stage—**seral community**—follows another it usually creates the conditions for its own replacement until a mature community or **climax** is reached which is theoretically self-perpetuating and in stable equilibrium with its habitat provided that the habitat itself remains stable. This, at any rate, is the classical view of succession.

There is a plethora of terms to describe seres of various kinds on different surfaces. Where plants begin their succession on virgin territory—rock slides, sand dune, scree, pit heap, etc.—the process is termed a primary succession or prisere. These may be subdivided into lithoseres on rock surfaces, psammoseres on sand dunes, hydroseres in fresh-water bodies and haloseres on sea-shores or shrinking desert lakes. The former two may also be described as xeroseres as water shortage dominates their conditions at least initially. Where human interference prevents a sere from reaching climax, it is called a **broken succession** or **plagioclimax,** and as the inexorable pressure of increasing humanity bears ever more on virgin territory, it is likely that there will be few parts of the world where the plagioclimax is not the rule. Associated with the development of plagioclimax communities is the secondary succession which is never allowed to proceed to the full climax community because of frequent interruption. In many parts of the tropical world, the secondary succession with its rapid assimilation of materials and energy is the lynch-pin of successful agriculture, restoring the nutrient cycles after one variety or other of temporary agriculture has abandoned a cultivation site.

Although succession in plant communities has been studied for a considerable time, all the rules which govern the development of the community are by no means understood. However, observations on many communities have established some of the broad outlines of what takes place in most. These may be listed briefly as follows:

(a) The number of plant species increases rapidly during the early stages but tends to remain steady as climax is reached and may even decline (Odum 1963). There could be several reasons for this. In the early stages, for example on good soils, old fields, or burned forest with plenty of water, the surface is usually occupied by high producers needing lots of light. As shade increases, their place may be taken by less productive species which can tolerate shade, and these, in turn, may shade out many light-loving species. Conversely, on poor soils pioneers may be low producers, but as nutrients accumulate they may be eliminated by high producers, which themselves become subject to shade.

(b) Throughout the succession, the biomass total increases with time, and this may, and usually does, go on increasing after the maximum gross primary production has been achieved until it levels off at the climax of the succession. In the early stages of the succession gross primary production usually runs well ahead of community respiration, but at full maturity gross production, net production and biomass are in balance.

(c) The dynamic stability of a succession *decreases* with time. This is almost the exact opposite of the traditional view of succession which postulates that it is the final state which is the most stable. However, there is now fairly good evidence that stationary or near stationary late successional stages are very vulnerable to disturbance whereas earlier stages recover from disturbance much more quickly (Horn 1975). This has important implications for forest conservation management.

Theoretical models established on this basis suggest that 'patch-cutting', leaving intact forest around the patches, is likely to lead to more rapid regeneration to the 'stationary state' than wholesale cutting or even removal of all the commercially valuable trees.

Under certain circumstances reversal of succession is also possible, occurring particularly where farming systems overload the productive capacity of the ground. For example, in over-grazed grasslands, the grass cover is first reduced to be followed by the appearance of weeds and possibly soil erosion as the turf cover is finally destroyed.

The hypothesis that it is possible to predict the end-product of a succession as the 'climax community' has been very seductive during this century, and various workers have advanced models of the nature of climax communities. Some of the most important are listed in Table 3.

	Table 3 Some theories of the nature of climax vegetation	
Author	*General title*	*Nature of climax*
F. E. Clements	Monoclimax hypothesis	Climax represents the highest type of vegetation under its particular climate.
H. A. Gleason (1939)	Principle of species individuality	Species distribution is not determined by the inexorable climax-climate relationship. Each species is distributed individually according to its own response to climate and other habitat factors.
A. G. Tansley	Polyclimax hypothesis	Climate may well be the overall determinant of climax but each climax type would be expected to display several communities which could be regarded as climax-climatic, physiographic and edaphic (soil-related).
R. H. Whittaker (1962) (1970)	Prevailing climax	At any point the community will consist of overlapping populations distributed according to environmental gradients.
A. S. Watt (1947)	Cyclic hypothesis	In some vegetation mature 'climax' vegetation types may follow each other in rotation on the same site. (See also Aubréville 1938.)
H. S. Horn (1975)	Climax as a 'stationary state' in a statistical progression	Many-specied forests with mature trees will consist of community mosaics whose species composition depends on the distribution of statistical probabilities of success among the competing trees at initial colonisation or replacement. Thus the rules governing the ecology of plant succession are essentially statistical, at least in forests.

However, the anomalies which occur are so numerous that none of these general theories can do justice to all the variations of community/ habitat/time relationships found in nature. Thus a hypothesis that the climax community is a 'stationary state' of vegetation (Horn 1975) conflicts with that of Watt (1947) which suggests that some mature communities can follow each other on the same site, in rotation. Similarly Whittaker's model of the mature community as 'the prevailing climax', composed of the overlapping distribution of species' populations arranged along environmental gradients, is hard to match with the conditions of the tropical rain forests. Moreover, communities of this kind would be expected to change gradually over space into other kinds of prevailing climax, yet sharp boundaries are readily observable in nature. For example, the 1969 Royal Society/Royal Geographical Society joint expedition to the Matto Grosso in Brazil found very sharp boundaries between savanna and dry forest, closely related to a change in soil texture with few characteristics of continuous change across the boundary (Askew *et al.* 1970-1).

Because of the lack of general agreement about the theoretical basis of community behaviour, a number of different systems of analysis, classification and mapping have developed. Unlike topographic maps, however, where there is some general agreement about the data to be marked, vegetation maps are highly interpretive documents, often peculiar to the country of origin, the school of vegetational analysis, or even the individual worker developing his own ideas.

6.6 Community analysis and classification

The analytic and classificatory methods employed in vegetation study fall into three main types: physiognomic, floristic and ecological.

Physiognomic systems, i.e. those based on the appearance of vegetation, have a long history but have been criticised by some botanists as being too imprecise to be really useful. Moreover, there is often confusion about terminology from one system to another, particularly where workers have attempted simultaneously to classify both the vegetation and the controlling environmental factors. For example, the term 'tropical montane forest', which implies connections with climate and site, has been widely used in the literature, but it has been noted that the limits assigned to it on various criteria span a range of no less than 2600 m (Robbins 1968).

In spite of these difficulties, most ecologists and geographers are led to the inescapable conclusion that the appearance of vegetation reflects fairly faithfully the sum total of all the ecological factors of habitat, and physiognomic methods are therefore by no means redundant. Also, there are still many parts of the world—and these are legion in the tropics—where precise knowledge of the interactions which control the nature of communities is so lacking that physiognomic methods provide the primary means of access to the meaning of vegetational patterns (especially as they are particularly suited to aerial photographs).

The most sophisticated physiognomic system of vegetation analysis is associated with Küchler (1966) in the USA. It has been tested on different vegetation types around the world and sets out to overcome some of the disadvantages mentioned above. It is too complex to list here in its entirety, but it uses a standard terminology that is fairly unequivocal, does not require taxonomic knowledge and can be used in conjunction with maps of any scale, country or region. The system relies on letter and number symbols and letter combinations (formulae) to designate the various types of vegetation. This system is now widely taught and used in the USA where its speed and efficiency, even when used by workers with little botanic knowledge, have proved its utility time and again.

Floristic methods involving much more specialised knowledge have been widely employed in continental Europe especially and are particularly associated with the Swiss phytosociologist J. Braun-Blanquet. The methods are often described as the Zuricho-Montpellier system, from the botanical schools with which they are associated. They require very precise botanical information before the classification can begin and the ultimate aim is to produce a hierarchical ordering of all the community types of a region which is presented as a species table (see Note page 112). Although the methods have been criticised (Küchler 1967), the system has been successfully applied in both temperate and tropical lands. In the latter, however, attempts to apply the methods to lowland rain forests have met with little success (Longman and Jenik 1974: 74-5). Indeed, many authors deny that the concept of the plant association, which is the basis of the community typing at which floristic methods aim, is applicable in any case to lowland rain forests.

To avoid some of the problems associated with both purely botanic and purely physiognomic systems, an ecological-physiognomic system has been devised which is intended to provide an internationally acceptable system for vegetation classification. This has been adopted by Unesco (1973) but has not yet been universally accepted by all ecologists (Whitmore 1975). As with most systems relying on the appearance of vegetation, the basic unit of classification is the **formation,** i.e. a vegetation type distinguished by the unity of its structure, physiognomy and floristic composition. These may be grouped at a larger scale into **formation-types**. Thus within the tropical Far East at least thirteen distinct forest formations can be distinguished within the major formation-type of the tropical rain forests (Whitmore 1975: 121). The assumption is made in this classification (and others which resemble it) that the formation structure and physiognomy are ecologically significant and represent a response to habitat factors. As to precisely what habitat factors may be significant and how they interrelate in many forest types, especially tropical ones, there is as yet limited evidence. In complex vegetation it is probable that field observation will need to be supplemented by accurate model building and computer simulation of performance to reach adequate conclusions about the nature of habitat conditions and vegetational response (Parkhurst and Louks 1972).

Some American authors have preferred the term **ecosystem-type** to the term formation, implying a unity of all organisms within the ecosystem-types recognised. In European literature the terms **biocoenosis** (plants + animals + micro-organisms) and **biogeocoenosis** (biota + non-living habitat) are used in a broadly similar way. However, whatever the terms used and concepts implied, the maps drawn by authors who use these terms usually resemble closely the familiar formations and formation-types of physiognomic analyses. This is not particularly surprising, of course. As plant life dominates the working of an ecosystem because it is the main channel for the disposal of energy and materials within it, it might be expected to be used as the critical means of distinguishing between one ecosystem and another.

The advent of the computer has led to the widespread application of statistical techniques to vegetational classification (Harrison 1971). Broadly, the methods which have been developed or adapted for vegetational analysis originate in two different views of the plant community. The first view comprises those techniques which emphasise the continuous nature of vegetational variation; the second those which assume that it is possible to identify and classify distinct communities within vegetational types.

Collectively the statistical methods representing the former view are known as 'ordination techniques'. The term 'ordination' is defined as 'an arrangement of units in a uni- or multi-dimensional space' (Gittins 1968). Ordination embraces a variety of techniques, not all of them statistical. Thus, the familiar geographical transect is an 'ordination', as is the gradient analysis of Whittaker.

In vegetation study the units to be manipulated by the ordination technique are usually field data—species frequency, abundance, coverage, etc. or habitat factors—which are arranged in such a way that insights may be provided into the ecological processes that may have generated the observations. The assumption is made at the outset that a latent structure of some kind exists in the observations. Ordination techniques are particularly valuable when the number of variables is large so that connections between species distributions and environmental factors are difficult to see. The manipulations provide ways of arranging data so that those relationships between plant and environment that will repay closer study are revealed (see Figure 37). Once the insight to a possible causal factor or set of causal factors has been obtained, ecologists usually investigate further by controlled experiments or close field observations.

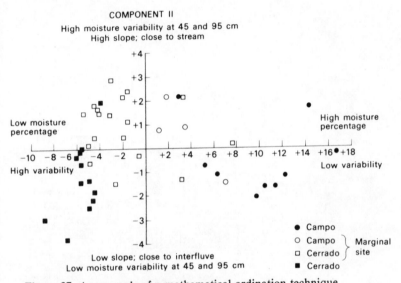

Figure 37. An example of a mathematical ordination technique (Principal Components Analysis). In this example twelve environmental variables from forty-three monthly-sampled sites in the north-eastern Matto Grosso have been distributed to the two axes Components I and II. These axes describe an ellipsoid scatter of points, each point representing a correlation between the variates studied. It can be noted in this example how the Component Analysis of the correlation matrix has differentiated significantly between the largely grassy vegetation (campo) and woody scrub vegetation (cerrado) (after Daultrey).

Statistical methods which attempt to analyse communities so as to yield classifications of floristic significance are exemplified by the work of the 'Southampton School' in Britain (Lambert and Dale 1964). These involve the collection of data by field sampling using the traditional quadrat as the sampling unit and arranging the data in a two-dimensional matrix in which all possible combinations of species pairs are compared. What the investigators are looking for is significant social groupings or sets containing species which tend to be associated together (see Figure 38).

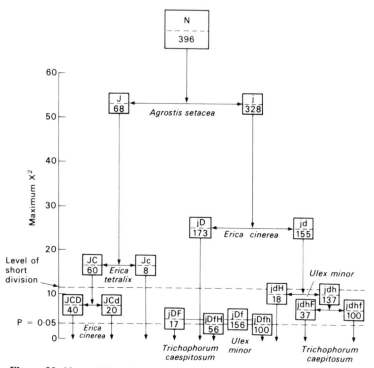

Figure 38. Normal Association Analysis to produce a hierarchy of ten species in the New Forest (after Lambert and Williams 1960). Data drawn from 360 quadrat sites. Species presence is indicated by capital letters, their absence by lower case letters.

The methods of Association Analysis have yielded significant results in the rather impoverished vegetation of the British Isles, but it is amongst the profusion of tropical vegetation that a method reveals its true mettle. The first application to tropical vegetation was not a success. In fact, so far away from ecological reality was it that one leading forest ecologist was forced to the conclusion that the more traditional physiognomic-structural analysis yielded better classificatory systems than the statistical methods (Webb *et al.* 1967). However, modification of the technique, especially in the matter of quadrat size, produced much more satisfactory results and led to the identification of what appeared to be coherent social groups of species within these complex forests (Webb *et al.* 1970).

The methods of analysis briefly outlined in this section are by no means diametrically opposed. Undoubtedly, if we were confined to a single technique in trying to understand the meaning of vegetational pattern, this would be quite inadequate. To have so many carefully developed and continuously refined analytic systems at our disposal is, in fact, a very healthy state of affairs.

6.7 The mapping of plant communities

Reflecting the variety of ideas as to how vegetation should be analysed and classified is the variety of ways in which it may be mapped. There is no single map of vegetation which can represent all the information relevent to its character or which can satisfy all the needs for information about it. This has been neatly demonstrated by tests devised by Küchler (1956) of mapping methods over the same piece of ground. Taking the varied needs of a soil scientist, a geographer in an underdeveloped country, an animal ecologist, a commercial beekeeper, a military strategist and a tax assessor (It is surprising just who needs to know about plant geography!), it was quite clear that the vegetation map which best fulfilled the needs of each would require a different scale, method of representation, and classification.

Although the diversity of needs for information about vegetation is very wide, most published maps result from the particular methods and needs of botanists, geographers and ecologists, and tend to reflect the priorities of each discipline. Thus, most maps may be described as physiognomic, floristic, regional or ecological, and may be drawn on small, medium or large scales. No matter into which category they fall, they all rely critically on the sensitivity and internal fidelity of the classification system adopted.

From the nineteenth century down to the present day, physiognomic maps, because of their utility for a variety of purposes, have constantly been published. On the whole, their relative simplicity makes them particularly suitable for land planning, forestry, conservation, agriculture, military planning and these days for rapid aerial survey. They have been published on a variety of scales although the use of medium scales is rare as it tends to make the maps too small for synthesis and too large to remain sufficiently accurate. Notable examples of physiognomic maps include that of the USA by Küchler (1964) and those by Gaussen in the *Atlas de France,* and there are many others in the various great world atlases.

In France the vegetation map has become a subtle and informative document carrying the maximum useful information about plant communities and their ecology. There are two main centres for production of these maps. One at Montpellier produces maps for the purposes of scientific research into the problems of vegetation, and these are usually of specified critical areas at a scale of 1/20 000 or larger. The production of these maps is integral to the work of a scientific institute, the 'Centre d'études phytosociologiques et écologiques', which carries out field survey, classification, and experimental investigation of the problems of vegetation. The other main centre, the 'Service de la Carte de la Végétation de la France' at Toulouse, is concerned with publishing maps of more general use at a scale of 1/200 000 based on the methods developed by H. Gaussen. These maps are much more empirical in scope and include information about environmental conditions as well as natural and agricultural vegetation, and have additional inset maps at 1/1 250 000 showing aerial coverage, administrative boundaries, land use, climax vegetation, soils, agricultural hazards and annual means of temperature and precipitation. There is also a wealth of marginal textual information. Altogether, the beautiful and subtle colour printing and layout make these a triumph of the cartographer's art and the maps are superb regional geographies in themselves. Deservedly they have become central documents in the planning process in France. The Toulouse centre has taken on the formidable task of mapping world vegetation at a scale of 1/1 000 000, and although this has meant a reduction in the numbers of colours and symbols, the maps, together with their inset and marginal information, are nevertheless indispensable to the modern study of world biogeography.

Note

Page 107

The basic unit of the Braun-Blanquet system is the plant association identified as a statistically homogeneous floristic grouping taken from a floristically homogeneous surface which has at least one characteristic species. The suffix '-etum' is added to the generic of the characteristic species. Thus if the characteristic species is the beech, the association is the Fagetum association. The associations are then grouped into higher units: alliance (suffix '-ion', e.g. Fagion alliance), order and class. At the association level any differential species which occur are designated by the suffix -etosum. For example, if it happened to be the grass *Festuca ovina,* the association would be *Fagetum festucetosum.* When all the plant associations and their hierarchical arrangement for a particular region have been completed, they are then presented as a species table with appropriate maps.

References and Further Reading

General

Boughey, A. S. 1968. *Ecology of populations.* New York and London: Macmillan.

*Daubenmire, R. F. 1968. *Plant communities.* New York: Harper & Row.

Dice, L. R. 1952. *Natural communities.* Ann Arbor: U. Michigan.

Gause, G. F. 1934. *The struggle for existence.* Reprinted 1964, London: Butterworth.

Gaussen, H. 1960. *Méthodes de la cartographie de végétation.* Paris: Cntr. Nat. de la Rech. Sci.

Grieg-Smith, P. 1964. *Quantitative plant ecology,* 2nd edn. London: Butterworth.

*Kershaw, K. A. 1974. *Quantitative and dynamic plant ecology,* 2nd edn. London: Edward Arnold. (Probably the best practical guide available with good accounts of statistical methods.)

Küchler, A. W. 1964. *The potential natural vegetation of the coterminous United States.* New York: Am. Geog. Soc. Sp. Publ. 36.

Küchler, A. W. 1967. *Vegetation mapping.* New York: Ronald Press.

Lemée, G. 1967. *Précis de biogéographie.* Paris: Masson. (Good summary of Braun-Blanquet methods.)

Longman, K. A. and J. Jenik 1974. *Tropical forest and its environment.* London: Longman.

Unesco 1973. *International classification and mapping of vegetation.* Paris: Unesco.

Watts, D. 1971. *Principles of biogeography.* London and New York: McGraw-Hill.

Whitmore, T. C. 1975. *Tropical rain forest of the Far East.* London: Oxford U. Press.

**Whittaker, R. H. 1970. *Communities and ecosystems.* London and New York: Macmillan.

**Willis, A. J. 1973. *Introduction to plant ecology.* London: George Allen & Unwin. (A good general introductory text.)

The nature of communities

Aubréville, A. 1938. La forêt coloniale: les forêts de l'Afrique occidentale française. *Annls Acad. Sci. colon.* **9,** 1-45. (See Longman and Jenik 1974, 64-78 for summary.)

Askew, G. P., R. J. Moffat, R. F. Montgomery and P. L. Searle 1970-1. Soil and landscape in the north-east Matto Grosso. *Geog. J.* **136, 137,** 370-6. (An interesting case study of a savanna-seasonal forest area.)

*Federov, A. A. 1966. The structure of the tropical rain forest and speciation in the humid tropics. *J. Ecol.* **54,** 1-11. (An interesting counterbalance to the theory of speciation by niche differentiation.)

Gleason, H. A. 1939. The individualistic concept of the plant association. *Am. Midland Nat.* **21,** 92-110.

*Horn, H. S. 1975. Forest succession. *Sci. Am.* **232**(5), 90-8.

Langford, A. N. and M. F. Buell 1969. Integration, identity and stability in plant associations. In Craggs, J. B. (ed.) *Advances in ecological research.* New York: Academic Press.

Mellanby, K. 1968. The effects of some animals and birds on the regeneration of oak. *J. Appl. Ecol.* **5,** 359-66.

Parkhurst, D. F. and O. L. Louks 1972. Optimal leaf size in relation to environment. *J. Ecol.* **60,** 505-37.

*Richards, P. W. 1973. The tropical rain forests. *Sci. Am.* **229**(6), 58-69. (See also Longman and Jenik 1974, 68-75.)

Rowe, I. 1967. Phytogeographic zonation: an ecological appreciation. In Taylor, R. J. and R. A. Ludwig (eds) *The evolution of Canada's flora.* Toronto: U. Toronto Press.

*Watt, A. S. 1947. Pattern and process in the plant community. *J. Ecol.* **35,** 1-22.

Webb, L. J., J. G. Tracey and K. P. Haycock 1967. Report. *J. Appl. Ecol.* **4,** 13-25.

The analysis of plant communities

*Bray, J. R. and J. T. Curtis 1957. An ordination of upland forest communities of Wisconsin. *Ecol. Mono.* **27,** 325-49. (A classic application of the continuum approach.)

Chapman, S. B. (ed.) 1976. *Methods in plant ecology.* Oxford: Blackwell.

Dickinson, G., J. Mitchell and J. Tivy 1971. The application of phytosociological techniques to the geographical study of vegetation. *Scott. Geog. Mag.* **87(2), 83-103. (A straightforward introduction to some of the main techniques.)

Gauch, H. G. 1973. A qualitative evaluation of the Bray-Curtis ordination. *Ecol.* **54**(4), 829-36. (A critique of the ordination approach of the 'Wisconsin school'.)

*Gittins, R. 1968. The application of ordination techniques. In Rorison, I. (ed.) *Ecological aspects of the mineral nutrition of plants.* Oxford: Blackwell.

Harrison, C. M. 1971. Recent approaches to the description and analysis of vegetation. *Tr. Inst. Brit. Geogs.* **52, 113-27. (A clear account of the main methods.)

Lambert, J. M. and M. B. Dale 1964. The use of statistics in phytosociology. In Craggs, J. B. (ed.) *Advances in ecological research.* New York: Academic Press.

Robbins, R. G. 1968. The biogeography of tropical rain forest in south-east Asia. *Proc. Symp. Recent Advan. Tropic. Ecol.* Varanasi (2), 521-35.

Webb, L. G., J. G. Tracey, W. T. Williams and G. N. Lance 1967-70. Studies in the numerical analysis of Australian rain forest communities. *J. Ecol.* **55**(I), **56**(II), **57**(III) & (IV), **58**(V).

Whittaker, R. H. 1962. Classification of plant communities. *J. Ecol.* **47**, 83-101.

The mapping of plant communities

*Küchler, A. W. 1956. Classification and purpose in vegetation maps. *Geog. Rev.* **XLVI**, 155-67.

*Küchler, A. W. 1966. Analysing the physiognomy and structure of vegetation. *Ann. Assoc. Am. Geogs* **56**, 112-27.

Küchler, A. W. 1973. Problems in classifying and mapping vegetation for ecological regionalisation. *Ecol.* **54**(3), 512-23.

PART TWO

The Pattern of World Vegetation

Introduction

One of the most difficult problems in ecology and biogeography is the correlation between vegetation and climate. The workers who laid the foundations of plant geography assumed that, viewed on a large scale, the major vegetation unit (formation-type) represented the response of the plant kingdom to a major climatic type. Thus, for each climatic division recognised by the climatologists there should be an appropriate vegetation type. However, the vegetational anomalies which occur in broadly similar climatic types around the world are so numerous that, as one worker has said: 'It is pointless if not positively misleading to make generalisations about such relationships' (Eyre 1968, see p. 182).

Yet, as we have seen (Figure 28) the plant/soil system *is* limited by the atmospheric system, so it would be equally erroneous to suggest that there is no correlation at all. That vegetation varies 'in tune with climate on a continental scale' (Walter 1971) is undeniable. In trying to establish the notation of this tune there are three main aspects to the problem:

(a) the nature of climate itself, i.e. what exactly it is that we measure, and what its biological significance;
(b) the fact that each species makes its own response to climate within the framework of its genetic inheritance;
(c) the fact that the relations of vegetation to climate seem to be more than simply the sum of these individual responses.

Standard meteorological returns on which climatologists base their climatic definitions lack essential data of biological significance, especially the all-important potential evapo-transpiration (PE) figure, and it has not really been proved that this figure can be adequately derived from standard data. The best-known attempt is that of Thornthwaite (see Hare 1954) and there is some evidence that the PE climatic types defined by his methods can be loosely correlated with vegetation variations (Figure 39). However, in the tropical world the index seems altogether too coarse to relate to vegetational variation

119

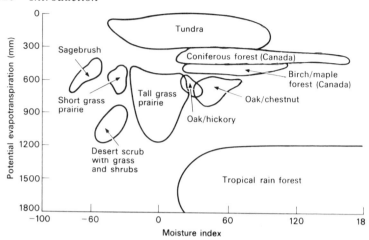

Figure 39. Relation between climatic moisture index, potential evapo-transpiration and natural vegetation. The areas enclose values from many observation sites and the natural vegetation at these sites is indicated. Note that the moisture index is obtained by the formula In = S/PE where S is the annual water surplus and PE the potential evapo-transpiration. (After Mather and Yoshioka 1968. Copyright Association of American Geographers.)

(Whitmore 1975). The method established by Penman (see p. 90) relies, in any case, on data not available from standard returns and its usefulness in relation to tropical vegetation is uncertain.

Even if it were possible to establish unequivocally a reliable index for PE, this would still not necessarily tell us about moisture availability to plants; there is also the factor of soil to consider. As Figure 28 makes clear, moisture is usually only available through the soil. In any two habitats with the same PE index but with soils of differing porosity and field capacity, the water relations of plants may be widely divergent. Thus in the example from the north-east Matto Grosso noted above (p. 106), the change in soil texture associated with the sharp boundary between dry forest and savanna probably operates to tip the balance in favour of one vegetation type or the other at a critical point in the moisture gradient.

A further difficulty in establishing the bioclimate is that of fire. This vital habitat factor is not easily quantified and, as we saw above (p. 75), its occurrence need not be frequent to produce important effects in vegetation.

A telling illustration of the complexity of species' response to the climatic environment has been provided by the work of Pigott on the stemless ground thistle (*Cirsuim acaulon*). It has been clearly shown that the failure of this plant to produce fertile fruit near the northern limit of its range in Britain is controlled by the complex interaction of *all* the elements of climate with which its distribution can be correlated. These elements—daily maximum temperatures, duration of sunshine, and rainfall during the summer months—act in concert to affect the physiology of this species' reproduction. Nor are the biotic effects of attack by slugs and fungus on its fruits by any means negligible factors in its inability to spread. As Pigott rightly points out, the neglect of the complexity of plant response to the equally complex variables which constitute climate is 'not easily excused'.

Where plants are aggregated together to compose a vegetation cover, a third variable is introduced. Many of the aspects of climate are profoundly influenced by the vegetation itself: wind speed, temperature, exposure of the ground to direct insolation, evapo-transpiration and so on. Although Gleason (1939) (see p. 105) convincingly argued the principle of species individuality, no species has, in fact, evolved in isolation from others. In relation to light, for example, there are few species which are indifferent to the shade cast by other plants. Plants of closed forests can usually only spread with the spread of forest. In the tropical lands, rain forest epiphyte species have evolved to take advantage of the microclimate within the tree crowns, and so on.

Finally, there is the whole question of Man's activities as a factor affecting climate. Where clearance and/or burning have occurred the climatic environment may be drastically altered. Thus the ground thistle mentioned above is usually found in calcareous pastures which are an artifact of Man to a large extent. The habitat conditions, climatic and otherwise, would have been radically different for this species before any clearances took place.

Because of these kinds of uncertainty many texts on vegetation geography contain 'poorly founded conclusions as to causation' (Pigott 1971). In truth we really know very little about the ways in which climate operates (a) to limit the distribution of individual species, and (b) to limit the distribution of individual vegetation types.

The real task in elucidating these relationships is to establish causative hypotheses which can be shown to be valid in the light of individual species' behaviour in relation to climate and the ways in which vegetation as a whole acts to modify the climatic environment.

Some of the difficulties noted above stem from the contrasting approaches of the ecologist and the geographer. Ecologists are concerned with precise field and experimental evidence on which to construct hypotheses to explain the observed facts: geographers tend to deal in the generalisation of very complex phenomena in order to discern the links by which various natural and human patterns on the Earth's surface interact to produce its varied characteristics. In historical terms, it was the approach of the geographer which dominated the view of world vegetation until comparatively recent times. As ecological research has proceeded, one of the most striking facts to emerge about generalised vegetation categories has been the diversity of ecosystem structure they conceal beneath their apparent uniformity. In one case after another the generalised categories of the geographers— savanna, temperate deciduous forest, tropical rain forest, etc.—dissolve to reveal intricate mosaics of minor ecosystems loosely contained in the overall framework determined by the macroclimatic characteristics.

Some of this mosaic pattern may be due to palaeoecological events, especially Quaternary climatic changes, and some to the activities of Man, but some appear to be inherent in the ecosystems themselves. In Part Two we shall look at the ways in which these and other factors dealt with in Part One interact to produce the pattern of the world's vegetation.

References and Further Reading

*Daubenmire, R. F. 1976. *Plants and environment,* 3rd edn. New York: Wiley.

Hare, F. J. 1954. *The evapo-transpiration problem.* Montreal: McGill University.

Penman, H. L. 1963. *Vegetation and hydrology.* Farnham Royal, Bucks: Comm. Bur. Agric.

**Pigott, C. D. 1971. The response of plants to climate and climatic change. In Perring, F. (ed.) *The flora of changing Britain.* London: Classey.

Schimper, A. F. W. 1903. *Plant geography upon a physiological basis,* trns. W. R. Fisher. London: Oxford U. Press. Reprinted 1960, New York: Harper & Row.

Walter, H. 1971. *Ecology of tropical and sub-tropical vegetation.* Edinburgh: Oliver & Boyd.

Whitmore, T. C. 1975. *Tropical rain forest of the Far East.* London: Oxford U. Press.

**Willis, A. J. 1973. *Introduction to plant ecology.* London: George Allen & Unwin.

7

Tropical forests

A. RAIN FORESTS

The tropical rain forests occupy some 400×10^6 ha of the Earth's surface, i.e. about one sixth of all the surface covered by broad-leaved vegetation. In spite of their extent and importance they are still probably the least understood in detail of all the world's major vegetation types. This is not surprising as they represent the most complex biocoenosis life has achieved with a high order of dynamic organisation and many unique features of morphology, life history and community interactions.

7.1 Structural characteristics and areal variations

The life-form groups which comprise the forest have been analysed by Richards (1952) as:

A. Autotrophs (i.e. with chlorophyll)
  (1) Mechanically independent plants—trees, treelets and herbs
  (2) Mechanically dependent plants—climbers, stranglers and parasitic and non-parasitic epiphytes

B. Heterotrophs (i.e. without chlorophyll)
  (1) Saprophytes ⎫ Fungi particularly
  (2) Parasites ⎭

Each of the groups (1) and (2) represents a partial life-form community (or **synusia**) within the forest. In any tract of forest the proportions of these synusiae will vary according to the **phase** of forest development. It is conventional to recognise three main phases of forest development: the gap phase, the building phase and the mature phase (Whitmore 1975). Although regeneration from the gap to the mature phase is continuous, the three phases can usually be identified in the forest physiognomy. The gap phase would be defined by seedling trees not exceeding 2.7 m and saplings <0.3 m girth, the building phase by pole-size young trees (0.3-0.9 m girth), and the mature phase by full-size trees. Thus the areal pattern of these forests is a mosaic of forest patches (Figure 14) at all stages from gap to mature. The species composition of these mosaics is usually highly varied, and gregarious dominant species are rare. Over large tracts of lowland forest the canopy will be composed of tree species not contributing more than one per cent to the total number of species. Family dominance, however, is more common. For example, in the lowlands of Malaya the Dipterocarpaceae may be said to be dominant (Whitmore 1975), although in these still-complex forests this dominance may be expressed by no more than the fact that the dipterocarpaceae provide perhaps 25 per cent of all mature tree species. This is in marked contrast to parts of Borneo where the dominant large dipterocarps may provide 90 per cent of all the emergent big trees. In the American tropical rain forests single-dominance is expressed especially by the legumes (*Epurua falcata* and *Mora* spp.) which may often form single-dominant stands on poorer soils. In Africa single-species dominance is associated with another legume (*Gilbertiodendron dewevrei*), again on unfavourable soils.

The areal pattern of tropical rain forest is also varied by the differing formations it encompasses. Some of these are shown schematically in Figure 40. As can be seen, differentiation into distinct formations also partially serves to differentiate the climatic and soil factors which influence their occurrence and distribution.

Floristically the four main regions (American, African, Indo-Malaysian and Australasian) of the tropical rain forests are quite distinct and there are only a few woody pantropical species and genera. Other non-tree plants, especially the epiphytes, are also regionally distinctive. Thus, in Indo-Malaysia the epiphytic synusia is dominated by the orchids, ferns, Asclepiadaceae and Rubiaceae. In America, in addition to orchids and ferns, the bromeliads and cacti are important (see Longman and Jenik 1974: 71). Moreover, within each of the major regions the floristic composition of the various formations alters gradually through space. In Indo-Malaysia, for example, the forest cover of the lands bordering the Sunda continental shelf has a rich dipterocarp flora (10 genera, 350 species) with many endemics, but north and eastwards this rapidly becomes impoverished.

The vertical structure of the forest has been described and drawn by many authors, usually by profile diagrams. But even now there is still some uncertainty as to whether or not the forest exhibits layering into identifiable strata as do the temperate forests. In one view there are—or should be—five recognisable above-ground strata:

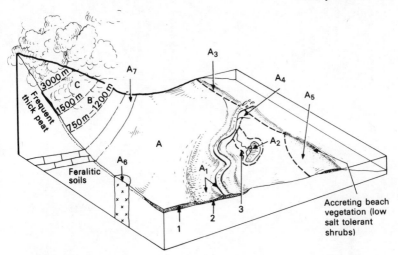

Figure 40. Formations within tropical evergreen rain forest vegetation type (formation nomenclature from Whitmore 1975). A—tropical lowland evergreen rain forest: A_1—fresh water swamp forest; A_2—peat swamp forest (extensive in Sumatra); A_3—mangrove forest; A_4—brackish water forest; A_5—heath forest over sands; A_6—forest over ultra-basic rocks occasional in South-East Asia; A_7—forest over limestone (rare in Africa). B—tropical lower montane rain forest. C—tropical upper montane rain forest. 1, 2 and 3 are common in Amazonian usage: (1) 'terra firma' with mesic forest; (2) 'várzea-forest' liable to inundation by 'white river' water; (3) 'igapo-forest' liable to inundation by 'black river' water. Alternative terms used by UNFAO classification: A—tropical ombrophilous forest (lowland type). B—tropical ombrophilous mountain forest. C—tropical ombrophilous cloud forest.

Upper tree layer >25 m (emergent trees with woody climbers and
    epiphytes)
Middle-tree layer 10-25 m
Lower tree layer 5-10 m
Shrub layer
Herb layer

plus 3 root layers from surface to 50 cm.

However, as Whitmore points out, this ignores the fact that the rain forest canopy is constantly changing, and any description that does not take account of the particular phase of forest development may be very misleading. Whitmore considers that a three-layer structure is a convenient working abstraction, encompassing the essential features of the building and mature phases, and this is adopted in Figure 41.

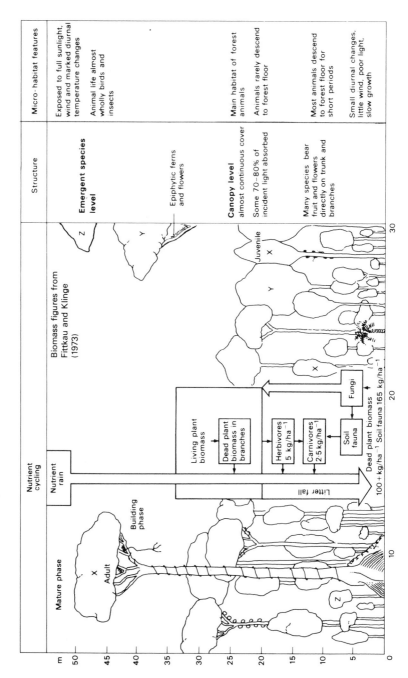

Structure	Micro-habitat features
Emergent species level	Exposed to full sunlight, wind and marked diurnal temperature changes
	Animal life almost wholly birds and insects
Epiphytic ferns and flowers	
Canopy level almost continuous cover	Main habitat of forest animals
Some 70–80% of incident light absorbed	Animals rarely descend to forest floor
Many species bear fruit and flowers directly on trunk and branches	Most animals descend to forest floor for short periods
	Small diurnal changes, little wind, poor light, slow growth

Biomass figures from Fittkau and Klinge (1973)

Nutrient cycling

Nutrient rain

Living plant biomass

Dead plant biomass in branches

Herbivores 5 kg/ha⁻¹

Carnivores 2·5 kg/ha⁻¹

Soil fauna

Fungi

Litter fall

Dead plant biomass 100 + kg/ha⁻¹ Soil fauna 165 kg/ha⁻¹

Mature phase

Building phase

Adult

Juvenile

Figure 41. Tropical lowland evergreen rain forest. (Composite profile adapted from examples in Richards (1952), Beard (1955), and Whitmore (1975).)

Whether specific plant associations occur related to generalised niches (see p. 99) or whether the rain forest is a haphazardly varying, single, unified community (see Ashton 1969, Richards 1969) is also still undecided. Probably the answer lies somewhere between these two views (Whitmore 1975). Discrete associations can certainly be identified. Most of the variation between these may be ascribed to a complex of factors including availability of flora, variation of topography and soil type, 'reproductive pressure' (i.e. in certain circumstances a variety of factors gives advantage to some species) and minor soil differences in uniform topographic types.

7.2 Climatic and other environmental relationships

Generally, the most important climatic characteristic to which the distribution of tropical rain forests can be related is that of rainfall. It is, however, quite impossible to give other than rather vague figures— >2000 mm mean annual rainfall, for example—as to how much rainfall is necessary before a rain forest vegetation can develop. As we saw above (Ch. 3.4), water supply to a plant depends on a number of variables, especially the balance of precipitation and evaporation. This is dependent in turn on topography, soil type, run-off and so on. To describe the tropical rain forest climate as hot and wet, with the assumption that precipitation always exceeds evaporation, is an oversimplification (see Figure 43 below). Although various attempts have been made specifically to design climatic extrapolations from standard data for rain forests (see Whitmore 1975: 43-51), none has been entirely successful. Important features of biological significance may be lost in the generalisation.

For example, over the last few decades detailed studies of the hot, wet forests have revealed surprising variations within the generalised climatic type. Thus, in South-East Asia annual periods of soil moisture deficit have been observed throughout the Malay peninsula apart from its southern tip and similar observations have been made elsewhere. Even at or near the Equator soil moisture deficits have been recorded. In the Congo, for example, the soil moisture has been observed to fall below permanent wilting percentage for the whole of the month of August (Longman and Jenik 1974: 35). These brief periods of moisture stress, seasonal or otherwise, have been recognised as important correlatives of flowering time in many species (Brunig 1971).

Other important bioclimatic features hidden in the average figures include the observations that

(a) moisture deficits are recorded more frequently at coastal stations than at inland stations;

(b) soil depth plays an important part in determining whether plants will suffer moisture deficit or not; in general, the deeper the soil the less chance of stress;

(c) the forest itself serves to control evaporation rates which have been observed to be less from open pans than from the vegetation (Douglas 1971);

(d) dry, sunny spells have effects on growth periodicity and may increase or decrease forest productivity according to site conditions. Sites with abundant nutrients produce significant growth, oligotrophic sites may decrease growth (Brunig 1971).

Rain forest plants must be adapted not only to these intermittent climatic variations but to the wide variation of diurnal climate inherent in the forest structure (see Figure 41 above). Thus, the plants of the forest are not hygrophytes but are generally mesophytes. Within the forest structure, the response to the vertical differentiation of micro-climate can be most easily observed in the leaves. On the whole the bigger, emergent trees have smaller leaves than those lower in the structure. Furthermore, the positioning of the leaves in the leaf mosaic of most trees seems to be a compromise between photosynthetic needs and the avoidance of moisture stress (see p. 64 on canopy shape). Indeed, in some species of emergents, for example *Piptadeniastrum africanum,* leaves markedly alter their positions in relation to the sun during the day, the movement being controlled by specialised **pulvini** (leaf joints).

The height which the forest reaches therefore depends on a number of variables—moisture availability, site conditions, any dry season, etc.—which determine the net production that can be sustained. Figure 42 illustrates the gradient of forest height observed through a transition between vegetation types in South America. It can be seen that the greatest heights obtain in intermediate conditions. Where rainfall is excessive (>3500 mm mean annual total), leaching of soil nutrients and reduced radiant energy due to cloud may significantly reduce forest production. In the Amazon Basin along the Rio Negro, low, open stands (Amazonian caatingas) occur in these conditions. Most rain forest emergents on mesotrophic sites can reach 50 m and in the most favourable conditions on eutrophic soils (for example, the upper parts of flood plains) the most luxuriant stands exceed 60 m. In general, forest with some seasonal drought is higher than continuously wet forest (Longman and Jenik 1974).

Many rain forest plants also show a clear response in growth rate to quite small temperature changes (in some cases 1°C or less). As the diurnal range of temperature at the emergent level can exceed 10°C

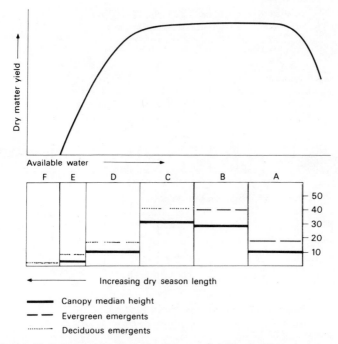

Figure 42. Relationship of yield, water availability and forest types (based on data from Ellenberg and Brunig 1971). A—Amazonian caatingas; B—tropical lowland evergreen rain forest; C—tropical evergreen seasonal forest; D—tree savanna/seasonal forest ecotone; E—tree savanna; F—desert scrub. Note that canopy heights shown are not the general rule for all tropical forests, e.g. in South-East Asia B would be generally higher in Dipterocarp forests.

and even at ground level may be more than 1°C, temperature change, however apparently insignificant, is an extremely important biological factor.

Growth in tropical rain forest species is not continuous. There is usually a definite period of shoot elongation which can be surprisingly short (in rubber and mango two weeks or less) and in many species a definite seasonality can be observed with the main peak of growth just after the 'dry' season (Medway 1972), as shown in Figure 43. The controls of growth rhythm and leaf 'flushing' appear to be very complex, however. In some species they seem to be a function of an internal 'biological clock', in others a response to variation in the external environment. One remarkable but little understood feature of leaf production is the spectacular colour of newly flushed leaves which may be pink, red or even blue. The lack of understanding of this

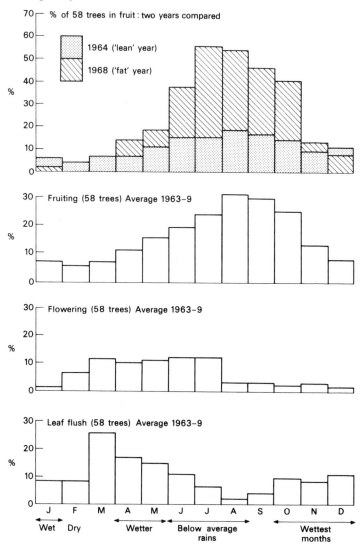

Figure 43. Seasonality in tropical rain forests (data from Medway 1972 based on Ulu Gombak forest, Malaya). Note the distinct relation of physiological activity to seasonal rainfall regime. Seasonal incidence of breeding (peak: March, April, May) and moulting (peak: August, September, October) in insectivorous birds are also related to the above rhythm as are variations from year to year of breeding success ('fat' years — high breeding success; 'lean' years — low breeding success).

common feature typifies much of our ignorance of the tropical rain forest. It is only in the last two decades that significant data have begun to clear the debris of misconception accumulated from the past and point to the ways in which much of our research must proceed.

7.3 Biomass, production and mineral cycling

Until recent years there were few reliable figures for the exact amounts of organic matter in undisturbed tropical rain forest. Various foresters' estimates existed, mainly from Africa and South-East Asia, and also figures of timber yields. In 1973, however, accurate figures based on careful study of the central Amazon rain forest near Manaos were released and showed a total plant biomass around $1100t/ha^{-1}$ for mature phase forest (Fittkau & Klinge 1973).

The most striking difference in the distribution of this biomass between these and other forests is the much greater preponderance of green assimilating organs (trees, ground flora and epiphytes). In the rain forests these represent 4-9 per cent of the total biomass, whereas in sub-tropical forests they are only approximately 2½ per cent and in temperate forests 1-2 per cent. (See also Lieth and Whittaker 1975. Note that in this text conclusions as to NPP in rain forest do not match this study or figures given in Whitmore 1975.)

Another surprising feature revealed by the Amazonian study is the low total biomass of animals which form only 0.02 per cent of the total living material. Few of these animals feed directly on green leaves (many of which are very bitter), but those which do, especially the leaf-cutter ants, have been shown to play an important part in mineral circulation (Lugo *et al.* 1973).

The huge mass of leaves carried by these formations provides a massive litter fall ($11t/ha^{-1}/an^{-1}$ in the Amazonian study area), yet soil reserves and surface accumulations are usually small except in lowland peat swamp forest and upper montane forest areas. In general, in the wet tropical lands litter is destroyed faster than it is supplied where mean temperatures are above 30°C. Between 25°C and 30°C litter supply and decomposition are equal. As humus turnover is one per cent per day there is little chance of its accumulating on the surface (Madge 1969). The most important decomposers of this litter are the fungi, and so effective are they that soil animals are forced to feed on them rather than the dead litter. Thus earthworms, in contrast to their role in mid-latitude soils, do little digging of the soil, so the organic upper horizon is usually sharply marked off from the mineral soil beneath.

In mature forest, although gross primary production rates may be high ($52.5t/ha^{-1}/an^{-1}$ in an Ivory Coast sample), there is evidence to suggest that net primary production is practically nil (Fittkau & Klinge 1973). Almost all the primary production is used in respiration. In the building phase, on the other hand, net production rates can be at least comparable to temperate vegetation. Thus, Whitmore (1975) compares Ivory Coast plantation figures to a forty-year-old beech plantation in Denmark and finds net production to be roughly equivalent. Rain forest managed for timber production, i.e. maintained in an artificial building phase state by reducing competition, can yield 3.6-$12t/ha^{-1}/an^{-1}$ of bole timber alone. Artificial plantations of conifers or certain *Eucalyptus* spp. can yield much higher than this (20-$35t/ha^{-1}/an^{-1}$).

The intense level of biological activity of the rain forest is maintained by a massive cycling of mineral elements, 3-4 times the amounts involved in other mature forests. By far the greatest element in circulation is silicon, with nitrogen, calcium and potassium next in importance. The key role in sustaining the massive biomass of the rain forest appears to be that of the nitrogen cycle. Apart from the massive litter circulation, any losses are made good by abundant nitrogen-fixing micro-organisms in the soil—bacteria, fungi and blue-green algae. These populations are themselves dependent on the ample energy-rich material supplied by the vegetation. In virgin forests, analyses indicate that soil nitrogen deficiency is rare. Many of the frequent legumes do not seem to form root nodule symbiosis with nitrogen-fixing bacteria, although one that does, *Cassuarina papuana,* has become important in managed crop rotations which are replacing haphazard shifting cultivation in the New Guinea highlands.

In spite of the huge circulation of nutrients, numerous observers have commented on the poverty of the soils from which they are drawn. With the comprehension of the massive biogeochemical cycles within these forests has come the realisation that much of the useful minerals are stored in the living biomass. For instance, one study (Nye and Greenland 1960) found that the ratio of soil nitrogen to biomass nitrogen was practically 1:3 and was 1:1 for most of the other important macro-nutrients except phosphorus. On their release from the vegetation by rain wash or litter fall or the death of plants, the minerals are taken up by the mass of fungi which infest the soil (see Figure 41) and returned to the plants very rapidly indeed. There is some difference of opinion as to whether mycorrhizae are as important as some authors have stressed, as they are certainly by no means universal to all tree species (Longman and Jenik 1974). So little escapes from the system of circulation that many of the streams originating in virgin forest have a composition *near that of distilled water.*

7.4 Soil types

The massive withdrawal of nutrients results in soils poor in the formation of stable organic and mineral compounds. At the same time, under conditions of high rainfall, high temperature and good drainage, the silica fraction of the soil becomes mobilised and leached downwards. These processes result in the most frequent soil type of these forests, the **feralites** (Buringh 1968), red soils which take their character from (a) the high content of iron and aluminium sesquioxides, i.e. oxides with two metal atoms to three of oxygen, and (b) simple clay minerals remaining, especially kaolinite. (Other names are also used for these soils—**nitosols** and **ferralsols** (FAO system), **oxisols** (7th Approx.) and **latosols**.)

Not all the soils below rain forest are feralites, however. Extensive tracts of podzols (**spodosols**) exist in Thailand, Borneo, Malaya, Guyana and the Rio Negro headwaters. These soils, usually developed over highly siliceous parent material (sands, quartzites, etc.), have lost their iron oxides through leaching and are very nutrient-deficient. Yet they can carry a dense vegetation often forming a distinct formation (heath forest), for example the 'keranga' and 'padang' of South-East Asia. These soils probably receive no nutrients from rock-weathering at all and the vegetation relies almost entirely on nutrients carried in by rainfall (Richards 1973). There are also soils described as brown earth types (**cambisols** — FAO; **tropepts** — 7th Approx.) which occur in south-west Nigeria and on recent lavas and seasonally-dry mountains in South-East Asia.

Most soils in the rain forests are easily damaged by forest removal and careless farming. On a small scale, clearances are soon colonised if sufficient primary forest remains, but on a larger scale, if rapid recolonisation is prevented, the upper layers can easily become waterlogged and begin to erode. Once the soil pores are filled with water, a process known as 'plinthisation' may set in. As the soil is now exposed to direct insolation, the high evaporation induces strong capillary action drawing up a stiff mottled clay which gradually fills the pore spaces. If this dries out an irreversible process sets in and a hard impervious duricrust forms. Once formed, this prevents—or appears to for a considerable time—the regeneration of the climax which is replaced by xerophytic herbs and shrubs and later simpler secondary forest of the kind found in many parts of South-East Asia.

Traditional farming practices based on small-scale clearances usually avoided the difficulties presented by the soil characteristics. Rapaport (1971) has shown that such systems can carry surprisingly high population densities given a sophisticated range of crops and the use of pigs. He shows that the wet forests of New Guinea can carry a population density of 124 people per square mile of agricultural land provided all the adults are engaged in farming. At the same time, the forest cover essential for forest regeneration can easily be conserved. Western agricultural practices, on the other hand, involving large-scale machinery and wholesale forest clearance, carry great dangers. The successful systems rely mainly on arboriculture for bananas, cacao, oil palm, and rubber which by careful management of the tree cover can be designed to avoid the processes of plinthisation and soil erosion. However, the wet tropical forests are at present under increasing attack for timber production, especially in South-East Asia, and in the Amazon for commercial farming in the wake of the Brazilian highway building programme. In the latter area, although the Brazilian government makes provision in its land sales for the preservation of forest in the sixty-mile plots each side of the road, the record of previous attempts to open up this great wilderness for speculative farming lends little confidence that this can be done without serious damage to the ecosystem.

The long, relatively undisturbed, evolution of this ecosystem-type has produced a battery of mechanisms encouraging stability of the system. Weeds and aliens are almost never found except where Man has intruded. Epidemics of fungal disease, insect larvae and pests seem to be rare in undisturbed forest (Richards 1973). Even the stability of the land surface and the control of erosion rates is part of the forest homeostasis. Little material is released to rivers, so they carry almost no load if they originate within the forests and consequently have almost no downcutting power. These are the 'clearwater' or 'blackwater' (peat-stained) rivers of the geologist. Any hard bands are crossed as rapids. (These are the habitats of a peculiar plant group, the Podostemaceae, with special adaptations for clinging to rocks.) At fault-scarps, waterfalls occur with no potholes at their foot nor any recession. In the Lower Amazon, for example, it has been found that some 80 per cent of the river load is derived from its Andean headwaters, not its huge lowland basin (see Douglas 1969).

As satellite photographs dramatically reveal, these forests are being removed at an ever-increasing rate. Richards (1973) has pointed out that the real cause for concern at their loss is not the damage it will do to the atmospheric oxygen balance or its alteration of the world's climate, but the fact that they will pass into history with their unique ecology almost totally uninvestigated. As Richards says, the tropical forest is a monument far older than the human species; it has many lessons to teach, if only we are willing to stop and learn.

B. SEASONAL TROPICAL FORESTS

The major formations which display these characteristics occur in South and Central America (Venezuela, Guyana, north-east Colombia, along a belt at the southern rim of the Amazon Basin and parallel to the coast in south-east Brazil), in South-East Asia, West and Central Africa, and Queensland.

The transition from evergreen rain forests to those with a greater number of seasonally deciduous species has been related to increasing length of dry season and decrease in rainfall (Figure 42). However, as shown in Figure 44, both the amount of deciduousness displayed and the number of species liable to shed their leaves vary according to soil type and altitude in any particular climatic regime (Webb 1968). Forests which can be ascribed to this type are much more variable in structure than the rain forests, as the increasing moisture stress has

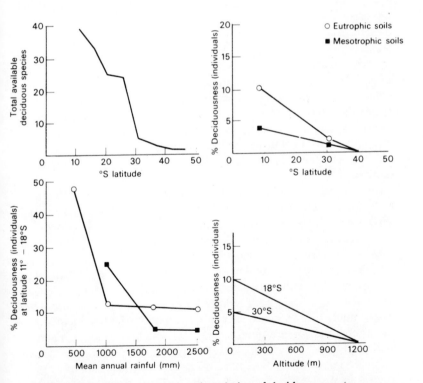

Figure 44. Diagrams illustrating the relation of deciduousness to various environmental factors in eastern Australian forests. (After Webb (1968). Copyright Ecological Society of America.)

provided greater opportunities for the development of adaptive features of life-history and life-form.

Unlike most of the rain forests, formations in this category often exhibit dominance by a few species, as in India and Burma where the sal (*Shorea robusta*), teak (*Tectona grandis*) and *Terminalia* species form major stands in the so-called 'monsoon forests'. The seasonal forests have almost a wintry appearance in the dry season but some observers have emphasised that it is in this season that many species flower. Those that flower—especially herbs—at the end of the wet season have been noted as responding to the seasonal variation in day-length (Hopkins 1968).

7.5 Biomass, production and mineral cycles

Living plant material in these forests appears to average about $205t/ha^{-1}$, but there are considerable differences between the wetter and drier ends of the climatic spectrum under which the forests develop. In contrast to the wet tropical forests, the dry season slows the destruction of litter (Figure 45), and greater quantities of dead material can accumulate to form a significant part of the total organic matter. The amount of dead material is also variable according to climate, there being usually a much more efficient destruction by soil organisms in the wetter types. Where the dry season is longer and the material resistant to attack, as with bamboo thickets in South-East Asia, the quantities of dead organic matter can be considerable. In one study in Thailand as much as $19\text{-}20t/ha^{-1}$ of dead bamboo was recorded.

The mineral circulation is said to have more in common with that of the temperate forests than with the hot wet forests (Rodin & Bazilevich 1967), especially as regards the proportions of the various minerals in circulation. However, there are few studies on which to base more than tentative conclusions.

The occurrence of fire in these forests introduces an ecological factor of considerable significance when compared with the hot wet forests. A dry season, even if only a few months long, combined with accumulated undecomposed litter, produces perfect conditions for the incidence of fire. These need not be annual to have marked effects. Occasional fires, possibly at long intervals, might well be sufficient to exclude the species of the hot wet forests. These have been shown by experiment in forest reserves in West Africa and India to be perfectly capable of colonising the wettest seasonal forest, but their thin bark and unprotected buds make them easy victims of fire.

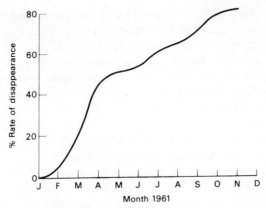

Decomposition of oak leaf discs in a
deciduous wood in Hertfordshire, England

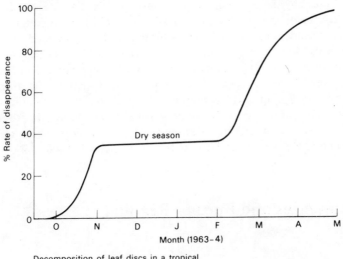

Decomposition of leaf discs in a tropical
seasonal forest, Ibadan, Nigeria (mean results
of four tree species)

Figure 45. Decomposition of leaf discs in tropical and temperate
forests. (After Edwards and Heath and Madge (1966).)

7.6 Soil conditions

The soils of these forests are not the feralites of the hot wet forests,
although they may often be red in colour. It seems that where
precipitation is below 1500-2000 mm per annum, feralites rarely form

(Ackroyd 1967) and the clay-humus content is much more stable with a greater reserve of available bases. Their frequent reddish colour is due to the mobilisation of iron which tends to form concretions. Although they need careful management, the more eutrophic types can be made very productive. In many parts of the world, especially South-East Asia and Africa, many of the forest areas have long been cleared or kept in some form of broken climax for a considerable time. Thus, it is impossible to get a true idea of the climax or of the original nature of the soil. In South America, the forest and soils are much less disturbed and it has been suggested that the frequent dominance by valuable species and the rapid regeneration rates—possibly about 100 years to maturity—make these lands more suited to forestry on a sustained yield basis than to agriculture (Budowski 1966). However, whether the world has time to work out proper management techniques for these soils is a moot point. Some of the most rapid increases in world population are associated with lands formerly or now partially dominated by these forests. South America, West Africa, the Indian sub-continent and South-East Asia all have rapidly expanding populations dependent on the capacity of these soils to supply food and raw materials for industry. Various world agencies and local institutions are engaged in many experiments to improve the agricultural practices, crop types, land management techniques and economic infrastructure in order to maximise the potential of the soil, but it is undoubtedly a race against time. After the great world food conference in Rome in 1974 it would seem that, for the time being, the surpluses of the soils of temperate lands must underwrite the deficiencies of these soils.

References and Further Reading

General

Buringh, P. 1968. *Introduction to the soils of the tropical and subtropical regions,* 2nd edn. Wageningen: Cen. Agric. Publ. & Doc.

Janzen, D. H. 1975. *Ecology of plants in the Tropics.* London: Edward Arnold.

Lieth, H. and R. H. Whittaker (eds) 1975. *Primary productivity of the biosphere.* Berlin: Springer Verlag.

**Longman, K. A. and J. Jenik 1974. *Tropical forest and its environment.* London: Longman.

Nye, P. H. and D. J. Greenland 1960. *The soils under shifting cultivation, Techn. Comm. 51.* Commonw. Bur. Soil Sci.

*Richards, P. W. 1952. *The tropical rain forest.* Cambridge: Cambridge U. Press.

Rodin, L. E. and N. I. Bazilevich 1967. *Production and mineral cycling in terrestrial vegetation.* Edinburgh: Oliver & Boyd.

Walter, H. 1971. *Ecology of tropical and subtropical vegetation.* Edinburgh: Oliver & Boyd.

**Whitmore, T. C. 1975. *Tropical rain forests of the Far East.* London: Oxford U. Press.

Vegetational characteristics

*Ashton, P. S. 1969. Speciation among the tropical forest trees: some deductions in the light of recent evidence. *Biol. J. Linn. Soc.* **1,** 155-96.

Beard, J. S. 1955. The classification of tropical American vegetation-types. *Ecol.* **36,** 89-100.

Budowski, G. 1966. Los bosques de los tropicos humedos de America. *Turialba* **16**(3), 287-95.

*Grubb, P. J. and T. C. Whitmore 1966. A comparison of montane and lowland rain forest in Ecuador. *J. Ecol.* **54,** 303-33.

Hopkins, B. 1968. Vegetation of the Olokomeji forest reserve, Nigeria, 5. *J. Ecol.* **56**(1), 97-115.

Medway, Lord 1972. Phenology of a tropical rain forest in Malaya. *Biol. J. Linn. Soc.* **4,** 117-46.

Richards, P. W. 1969. Speciation in the tropical rain forest and the concept of the niche. *Biol. J. Linn. Soc.* **1,** 149-53.

Richards, P. W. 1973. The tropical rain forest. *Sci. Am.* **229(6), 58-69.

Environmental relationships

*Brunig, E. F. 1971. On the ecological significance of drought in the equatorial wet evergreen forests of Sarawak (Borneo). In Flenley, J. R. (ed.) *The water relations of Malesian forests.* Hull: Dept. of geography, Hull University.

*Douglas, I. 1969. The efficiency of humid tropical denudation systems. *Trans Inst. Brit. Geogs* **46,** 1-17.

Douglas, I. 1971. Aspects of the water balance of catchments in the Main Range near Kuala Lumpur. In Flenley, J. R. (ed.) *The water relations of Malesian forests.* Hull: Dept. of geography, Hull University.

Webb, L. J. 1968. Environmental relationships of structural types of Australian rain forest vegetation. *Ecol.* **49(2), 296-311.

Production and mineral cycles

*Fittkau, E. J. and H. Klinge 1973. On biomass and trophic structures of the central Amazon rainforest ecosystem. *Biotropica* **5**(1), 2-14. (Very thorough study of undisturbed rain forest.)

*Lugo, A. E., E. G. Farnsworth, D. Pool, P. Jerez and G. Kaufman 1973. The impact of leaf cutter ants *Atta colombica* on the energy flow of a tropical wet forest. *Ecol.* **546,** 1292-302. (Interesting attempt to obtain data on a very complex relationship.)

Rapaport, R. A. 1971. The flow of energy in an agricultural ecosystem. *Sci. Am.* **224(3), 116-34.

Rapaport, R. A. 1976. Forests and man. *Ecologist* **6(7), 240-7.
(These papers by Rapaport provide a thorough study of the human ecology of rain forest inhabitants.)

Soil characteristics

Ackroyd, L. W. 1967. Formation and properties of concretionary and non-concretionary soils in western Nigeria. *Proc. 4th Reg. Conf. Soil Mech. & Found. Eng.,* Cape Town **1,** 47-51.

Madge, D. S. 1966. How leaf litter disappears. *New Sci.* **32, 113-5.

Madge, D. S. 1969. Litter disappearance in forest and savanna. *Pedobiologica* **9,** 288-99.

Young, A. 1974. Some aspects of tropical soils. *Geography* **59**(3), 233-9.

**Lowe-McConnell, R. H. (ed.) 1970. *Speciation in tropical environments.* New York and London: Academic Press. (This volume is compiled from nos 1 and 2 of the *Biological Journal* of the Linnaean Society. It provides an overall view of some of the main controversies in tropical botany.)

8

Tropical formations with conspicuous grasslands: savannas

There are a number of definitions of the word 'savanna', which in origin is an Amerindian term. In Europe, it is usually defined as a tropical grass-dominated formation with a greater or lesser proportion of open woody vegetation and associated trees. In the USA the word savanna is not confined in its usage to tropical formations. These rather vague definitions have challenged a number of authors to define the vegetation-type more precisely. Walter (1971) excludes formations that are not natural in origin. However, others make no such distinction. As research has proceeded within these lands, it has become increasingly clear that their ecology—superficially similar from place to place when judged by appearance—is very complex. Indeed, no better example of the complex, intricate mosaic patterning of communities can be found than those which exist in the savannas.

Savannas exist in *all* tropical climatic regions and are neighbours to practically all the tropical formation-types. They cover some 20 per cent of the world's land surface and are found in both mountain and lowland areas over a variety of soils. With such a diversity of habitats it is hardly surprising that their origins and ecosystem dynamics should be the subject of active controversy.

8.1 Physiognomic and botanical character

Floristically, the herbaceous layers tend to be strongly dominated at any particular locality by one or two species of various genera (*Andropogon, Loudetia, Hyparhenia, Pennisetum, Stipagrostis*, and in Australia *Thelmeda, Astrebla* and *Triodia*). The height range is from the several metres of Africa elephant grass (*Pennisetum purpureum*) through the habitual one metre or so to the low sparse grasses of areas with a long dry season. All the grasses are xeromorphic and many have rhizomes (underground stems) and densely-tufted aerial parts.

141

The shrub and tree strata are diverse and contain numerous species which display similar morphology. Their rooting is strong and may attain great depth, and their crowns are often flattened with small, hard, dry leaves shed during the dry season. Some, such as *Acacia Faidherbia albida* of the African Sudan-zone savannas, retain their leaves and many (except in Australia) are thorny. Most have thick bark and bud-scales which preserve them from fire and desiccation. Both trees and shrubs produce enormous numbers of seeds. *Acacia karoo*, for example, may release 20 000, 90 per cent of them fertile, most of which are eaten by termites, destroyed by fire or fail to establish themselves after germination.

The subject of competition between grasses and trees is complex and the reader will find various opinions held by different authors. Walter, for example, believes that they are direct competitors, especially for water at the end of the dry season. As grasses can withstand considerably greater drought than trees they should be more successful on fine soils where their dense rooting can more easily command their soil reserves. On stony soils, on the other hand, trees should be more successful as their deep roots can better exploit reserves at depth. Thus, on stony soils Walter envisages trees as being the true dominants with grasses only a tolerated part of the community. He presents the following scheme as indicating the main plant/climate/soil inter-relationships in summer rain areas (i.e. most of the savannas):

(a) stony soils dominated by woody plants;
(b) fine soils (250-500 mm mean annual precipitation) dominated by grasses with some woody plants;
(c) fine soils (over 500 mm mean annual) associated with savanna woodlands;
(d) fine soils (100-250 mm mean annual) dominated by pure grasslands.

The view that grasses and trees are direct competitors may well be valid. However, there is considerable evidence to show that grasses and trees can co-exist to form stable communities under a variety of climates and on many soil types. Moreover, the patterns of grass/tree associations have been very convincingly related to the mosaic of the soil/geomorphological units inherited from Tertiary times and modified profoundly during the last two million years.

8.2 Community patterns

The mosaic has been investigated in all three southern continents (Morrison *et al.* 1948, Bunting and Lea 1962, Cole 1960, 1965), and has been adopted by the Commonwealth Scientific Research Institute for the purposes of mapping and description, especially in Australia.

Cole identifies four major soil/vegetation units as successors to the widespread plains destroyed during the Pleistocene. These are set out schematically in Figure 46.

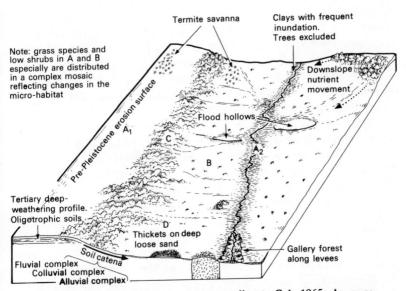

Figure 46. Savanna landscape types according to Cole 1965. A—pure grassland: A_1—grassland on plateau watersheds (low nutrient soils), A_2— valley floors with clay soils; B—savanna parkland (frequently located on eutrophic or mesotrophic soils); C—savanna woodlands (grass stratum present although tree crowns contiguous) generally associated with soils on old erosion surfaces; D—low tree and shrub savanna often on most recent skeletal soils.

Within each of the major associations are a number of minor associations reflecting differences in microclimate, relief and soil texture and mineral status. The minor associations may contain numerous tree species but only one or two can be regarded as dominants. The distribution of herbaceous species is usually independent of that of trees, and further intricacy is added to the local patterns by the effect of flooding (which can occur frequently on a variety of surfaces), the presence of flattened termite mounds ('termite savanna'), and the presence of steep slopes where vegetation often reflects the pattern of the soil catena from top to bottom (see p. 90).

The two approaches to the problem of savanna community patterns, the strictly ecological one of Walter and the biogeographical approach of Cole, present an interesting contrast. As yet there is no easy resolution of these theories.

One intriguing sidelight is cast on this controversy by the vegetation of south-west Australia. Here there are extensive 'fossil feralite' soils capping the highest erosion surfaces which carry a vegetation dominated by jarra and karri (*Eucalyptus marginata* and *E. calophylla*), essentially like savanna woodlands. This vegetation has been described as a 'relict ecosystem' that originated under a more tropical climatic regime with summer rainfall and which survives under a warm temperate winter rainfall regime. Its persistence has been ascribed to the extremely base-deficient soils which have kept out plants better adapted to the climate.

There are many areas around the world that can be classed physiognomically as savannas but are not easily placed in either system above. In Nicaragua, Colombia, Borneo, West Africa and the Philippines, extensive savannas are maintained as an anthropogenic fire-climax. In fact, in most of the savanna lands it is difficult to distinguish easily between 'natural' and 'Man-induced' formations, especially as fire is in any case a normal feature of the ecosystem. This is well illustrated by the grasslands of northern South America. Pollen diagrams from the Llanos Orientales of Colombia and the Rupununi Savannas of Guyana show that, even before Man appeared, frequently-fired, open, grass-covered plains existed in association with dry periods. However, during pluvial spells, for example in late glacial times, the same areas were dominated by dry forest or closed savanna woodlands which could survive the less frequent fires (Wijmstra and Van der Hammen 1966).

8.3 Biomass, productivity and mineral cycling

Because of the varied nature of the savanna lands it is impossible to give precise figures of the living biomass. In wooded vegetation it can be well into the range of closed forest—as much as 150t/ha^{-1}—but sparse grassland without trees may carry as little as 2t/ha^{-1}, i.e. below the mean biomass of desert scrubland. The gross primary productivity similarly shows a wide range between 2 and 20t/ha^{-1}/an^{-1}. In grassland there is a very direct relationship between this productivity and rainfall. Mean figures for grassy formations in Namibia indicate a rate of 1t/ha^{-1}/an^{-1} per 100 mm mean annual rainfall (Walter 1971). However, in woody vegetation this relationship is far from clear as many of the woody formations are associated with nutrient-poor soils which limit production.

The few studies of mineral cycling available are insufficient to give any more than a generalised picture. There is agreement, however, that the nutrient circulated in greatest quantity is silicon in both grasses and woody plants. So high is its concentration in some of the grasses that clots of amorphous silica (**phytolitharia**) may be formed in the leaves. The amounts of nitrogen returned to the soil annually are also high and are in fact not far below those quoted for tropical rain forest.

The processes of litter decomposition are mainly seasonal but there is not a lot of evidence as to how they take place. Both humification and mineralisation by soil animals are rapid during the wet season, but observers have noted that the soil fauna are likely to show the same behaviour as that in the rain forests, i.e. they feed on fungi which act as the main decomposers. In many savannas, the animals which play the most important role in contributing to litter decomposition are the termites. Whether earthworms contribute much to the process is doubtful (Madge 1969). There seems, in any case, to be a mutually exclusive ecological relationship between earthworms and termites.

The biomass of soil animals and vertebrates can be very large. In East African thorn-bush savannas, 250 kg/ha^{-1} of soil animals has been recorded, whilst figures for vertebrate herbivores in East and Central African savannas range between 100 and 300 kg/ha^{-1}. Careful studies in the Ngorongoro crater show how so many species in such bulk can be supported (Figure 47). The different game animals each have a particular ecological niche in the way they use the vegetation, and such is the mosaic of plant associations that the animals are presented with palatable food in all seasons (Anderson and Herlocker 1973). Whether big game favours grasses at the expense of trees is undecided. Probably of greater importance in many savannas are the termites which compete with game, cattle and Man for food and have considerable effects on the soil (Pullan 1974).

8.4 Soils and agriculture

The substitution of agricultural for natural ecosystems in the savannas is a distinctly hazardous operation in view of the variable nature of the soils, their frequent poverty, their highly variable rainfall from year to year, and the many pests and diseases that they harbour. The valley grasslands and parklands can be made reasonably productive, especially with irrigation, but watershed sites and low shrub and savanna woodlands are nutrient-poor and do not take kindly to the

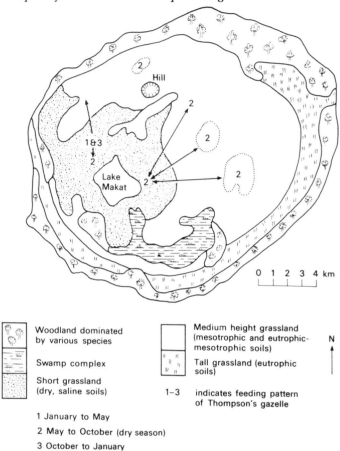

Figure 47. Generalised vegetation pattern of Ngorongoro crater, Tanzania. Dotted areas indicate depressions with *Pennisetum/ Andropogon/Cynodon* grassland particularly favoured by Thompson's gazelle. (After Anderson and Herlocker (1973).)

plough. Extensive mechanised farming carries with it many risks, for example soil erosion, soil compaction, rapid exhaustion of nutrients, re-invasion by suckers and the encouragement of pests and aliens where the ecological balance has been disturbed. The peasant subsistence farming of Africa and South America usually avoids these difficulties by the use of shifting cultivation or 'bush-fallowing'.

The most important commercial agricultural activity in all three southern continents, however, is cattle raising. As with plough farming, the variable potential of the soil/vegetation mosaic is the most important determinant of forage yield. The mosaic is, nevertheless, open to improvement and management as the experiments at the Kongwa ranch in Tanzania have shown. The establishment of exotic pasture grasses on suitable sites allows a carrying capacity as high as one beast per two acres, a far cry from the one beast per several square miles of the Australian outback.

By far the greatest limitation on increasing cattle yield is disease, especially rinderpest, sleeping sickness and foot-and-mouth disease. Although, in Africa, local veterinary methods can be very effective, as with the practices of the Bamangwato peoples of Botswana, the problems of cattle production can only be solved by widespread and effective preventative veterinary medicine allied to land management research.

As yet the inexorable rise in human numbers in these lands is less immediately pressing than in other parts of the world. Nevertheless, within the next few decades, most of them will need to be brought into a managed system, especially in Africa. Given adequate knowledge and the political will to apply it, there should be no reason why these lands could not support greater numbers and still continue to provide for the indigenous flora and fauna which, in Africa above all, constitute one of Man's most spectacular heritages.

References and Further Reading

General

**Cole, M. M. 1965. *Biogeography in the service of man.* London: Bedford College.

Harris, D. R. 1974. Tropical vegetation: an outline and some misconceptions. *Geography* 59(3), 240-51.

**Walter, H. 1971. *Ecology of tropical and subtropical vegetation.* Edinburgh: Oliver & Boyd.

Specific aspects

*Anderson, G. D. and D. J. Herlocker 1973. Soil factors affecting the distribution of the vegetation types and their utilization by wild animals in the Ngorongoro Crater, Tanzania. *J. Ecol.* 61(3), 627-53. (A brilliant exposition of the interrelationships of soil, vegetation and animal life in relatively undisturbed savanna.)

Blydenstein, J. 1967. Tropical savanna vegetation of the Llanos of Colombia. *Ecol.* 48, 1-17.

148 *Tropical formations with conspicuous grasslands: savannas*

*Bunting, A. H. and J. D. Lea 1962. The soils and vegetation of the East Central Sudan. *J. Ecol.* **50**(3), 529-58.

Cole, M. M. 1960. Cerada, caatinga and pantanal: the distribution and origin of savanna vegetation in Brazil. *Geog. J.* **CXXVI**(2), 168-79.

*Lamb, H. 1973. Is the Earth's climate changing? *Ecologist* **4**(1), 10-15.

Madge, D. S. 1969. Litter disappearance in forest and savanna. *Pedobiologica* **9**, 288-99.

Morrison, C. G. T., A. C. Hoyle and J. F. Hope-Simpson 1948. Tropical soil-vegetation catenas and mosaics. *J. Ecol.* **36, 1-84. (Classic exposition of the biogeographical approach to savanna vegetation.)

*Pullan, R. A. 1974. Biogeographical studies and agricultural development in Zambia. *Geography* **59**(4), 309-22.

Wijmstra, T. A. and T. Van der Hammen 1966. Palynological data on the history of tropical savannas in northern South America. *Leidse Geologische Medelingen* **38**, 71-91.

9

The vegetation of arid lands

9.1 Locations and characteristics

According to Unesco's Programme for the Arid Lands, areas which suffer from lack of water throughout the longer part of the year cover some 35 per cent of the world's land surface (Figure 48). However, to define them precisely is not easy. Rainfall alone cannot be used as they cover tropical and temperate lands with very different evaporation rates. The various indirect criteria which have been proposed—the presence of lakes with no outflow, intermittent streams, the presence of saline areas—are not particularly satisfactory either. Moreover, at least four kinds of arid climate can be recognised:

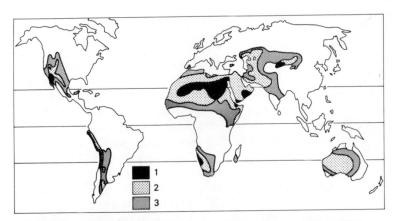

Figure 48. The world's arid lands according to Unesco. (1) extremely arid lands (true 'deserts'); (2) arid lands (regions where aridity poses serious limitations to farming and normally only grazing is possible); (3) semi-arid lands (regions where plough farming is possible but requiring specialised techniques for water conservation or irrigation). Note that the vegetation of the last regions would usually be classified into a number of formation-types, for example savanna, steppe or seasonal tropical forest.

(a) winter rain and summer drought climates (parts of the Middle East, the American south-west and the northern Sahara);

(b) summer rain and winter drought types (central Australia, the southern Sahara, the northern Atacama);

(c) climates with two rainy periods or no definite rainy period (south-west Africa);

(d) regions of extreme aridity with long runs of rain-free years.

The Unesco definition allows only the last category to be classed as true 'desert' and although the word may be used in many parts of the world for generally arid lands many of these 'deserts' would be excluded from this definition. Thus, it can be seen from Figure 48 that no true deserts are recognised in Australia although the map of that continent is liberally scattered with named 'deserts'.

In spite of low rainfalls and high evaporation rates in all the areas marked in Figure 48 plants manage to survive in nearly all these lands in one form or another. A primary distinction which can be made amongst these plants is between those which survive as permanently-rooted species, the perennials, and those which survive adverse conditions as seeds. With the exception of those plants that rely on dewfall or frequent fog, most perennials have water within range of their roots. The annuals, on the other hand, complete their cycle within a very short time while water is available.

The community patterns displayed by perennial species are closely related to available water. Thus in the transition from wet to dry areas the rule is usually for plants to become increasingly contagious in their distributions. They would be found clustering in hollows or other sites where run-off is concentrated. Where the land is very flat with few hollows or runnels, flood-sheets may occur (slopes as low as 1 in 2000 are sufficient) and perennial species usually increase in frequency in the direction of flood movement. This results from the fact that the longer the sheet covers the ground surface the more water is able to penetrate the ground.

The annual species, with their rapid germination, growth and flowering, are not exempt from the rules of plant community development simply because their season is so short. For example, germination is often temperature-dependent and if rain comes at the wrong time of the year many species will remain dormant whilst others may germinate. Once germination has taken place there is often intense competition for scarce resources. Thus, if the upper soil layers are liable to dry out quickly the more xeromorphic types are liable to dominate, but if water is relatively freely available the less xeromorphic plants tend to be more important. Also if there are dry spells within any rainy season, those annuals able to resist the drought are likely to be the survivors.

9.2 Biomass, production and mineral cycles

The distribution of biomass in arid conditions is mainly underground (see Figure 49), to the extent in some communities of more than 80 per cent of the living material. The green parts can be as little as one per cent of the total in shrub communities although it would be a higher proportion where annuals are conspicuous. Annual net primary production is very closely related to rainfall. Below a mean annual rainfall of 80 mm, the mean rate of fall in production per 10 mm of rain is about 200 $kg/ha^{-1}/an^{-1}$ (Walter 1971). However, if plants are tapping imported sources of water underground via an aquifer—the classic Saharan oasis—or along a river, production levels can be quite high.

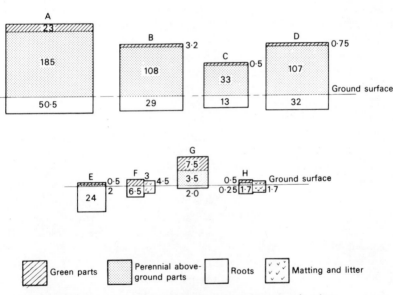

Figure 49. Amounts and distributions of organic matter in plant biomass in selected vegetation types (logarithmic scale) (data from Rodin and Bazilevich (1967)). A—tropical lowland rain forest (mean value); B—hornbeam forest, Crimea, USSR; C—42-year-old birch forest, Great Britain; D—green moss spruce forest (73 years old), Velikiye Luki Province, USSR; E—woody desert vegetation, old Anu-Dar'ya delta, USSR; F—steppe meadow, western Siberia; G—bog moss (Sphagnum) upland moor, Novosibirsk Province, USSR; H—Arctic tundra communities (mean value). All figures in tonnes/ha^{-1}.

The **tugai** formation of Central Asia (forest-shrub-grass communities along rivers) have levels of production comparable with those of temperate deciduous forest. In extreme deserts, net primary production is usually less than 50 kg/ha^{-1}/an^{-1}.

It is difficult to give a generalised picture of mineral cycles in arid regions as these are so varied in type. In any case, the bulk of elements is returned to the soil via root material (80-90 per cent) which is difficult to measure. Apart from tugai, it would seem that nitrogen is prominent amongst the elements returned in most communities. However, where succulents (cactus, etc.) form a prominent feature of the permanent vegetation there are considerable amounts of chlorine, sodium and sulphur returned to the soil.

9.3 Soils of the arid lands

In spite of their low production, the soil-forming processes of deserts are just as dependent as in the better-watered lands on the cycling of mineral elements. In very arid areas (category (d) above) there are rarely any surface mantles which can be described as true soils. However, if regular seasonal rains occur there are usually soil mantles.

A major distinction that can be made is between those soils which are freely drained and those which are not. Most of the former type under scrub are light brown or grey with little material supplied by the vegetation which has poor humification and mineralisation characteristics. Where grasses are dominant, these soils have a higher humus content and often a reasonable crumb structure with a significant biomass of soil animals. Most freely drained arid land soils contain a gypsum or lime horizon which may emerge locally with greater aridity at the surface as 'caliche' (North America) or 'tosca' (Argentina).

On land with impeded drainage, semi-shrub and shrub communities with annual herbs are typically underlain by **sierozems.** These usually exhibit sod formation and are constantly restructured by the roots of annual species, rodents and insects. Litter is usually rapidly mineralised but only small amounts of humus are produced. In sub-tropical deserts, iron, aluminium and calcium are often mobilised and assimilated and cycled by the plants, and their reaction with the soil acids (especially iron) results in ferruginisation, producing the characteristic reddish-pink of the soils (Rodin and Bazilevich 1967).

Common salt often accumulates in soils in arid zones. In some areas, for example Lake Eyre and around the Caspian Sea, the salt has been derived from the evaporation of sea water; but in other lands, as in Central Asia or central Australia or the western USA, the salt has been imported from the sea via rainfall.

The very arid lands with their scarce skeletal soils are usually associated with a very few characteristic species, depending on the part of the world in which they are located. In the Sahara, for example, the dune lands (erg) are dominated by the drinn grass (*Aristida pungens*) and a legume, the retem (*Retama taetam*). In the rock and stony deserts (hamada and reg) there are large tracts with no vegetation at all, plants being confined to soil pockets where grasses of the genus *Aristida* and woody plants belonging to the Chenopodiaceae can survive. Here plants may have to rely almost entirely on dewfall. In slightly saline conditions species of the *Tamarix* and *Nitaria* occur. The fact that plants can survive at all in some of these locations is testimony to the persistence and durability of living matter against almost impossible odds. In some of the most difficult locations of all where manganese coats the rocks, giving them a dark, heat-absorbing surface, temperatures at the hottest parts of the day reach well beyond the lethal temperature of living matter, yet plants and animals survive even here.

9.4 Man and the arid lands

Although the world's deserts support very low populations, the scarcity of their vegetation and its low productivity make them very liable to damage by any mishandling. In the Sahara, for example, the grazing herds of the nomads and their cutting of wood for fuel have easily demonstrable ecological effects which are greater than simply the removal of plants. Where plant cover is markedly reduced it has the effect of lowering the water table and increasing the soil drought so that once-fertile oases may be made completely useless. At the southern edge of the Sahara where it impinges on the semi-arid region of the Sahel zone, the practice of burning is common as a method of land clearance. The loss of soil moisture once the woody plants are killed may have resulted in a measurable advance of the desert.

For some there has been considerable speculation as to how far and how fast the arid zone is extending and what role Man has played in this. There is some evidence to suggest that the process is part of a world-wide change of climate which began some decades ago, but whether Man has accelerated a process that was inevitable anyway is not easy to say. Some American authors have speculated very freely on this subject, but the amount of hard evidence is limited.

Undoubtedly the deserts can be made to bloom with irrigation but it is essential to remove the salt in some way before the land can be used. If the salt is simply washed out by excess irrigation water, the soil may lose its structure and become poisonously alkaline. The Punjab, and the Imperial Valley of California, have suffered in this way. Techniques developed to avoid this include heavy dressings of lime or gypsum before irrigation water is introduced. As the sodium is lost, its place is taken in the clay-humus complex by calcium, improving its structure considerably. Many authors have seen the deserts with their high available energy as providing the potential new land for cultivation of food crops which the world will need to support its rapidly increasing population. Various proposals have been put forward involving the desalination of sea water (using solar energy, atomic energy or freely available oil supplies), the importation of irrigation water from surrounding better-watered lands or the exploitation of underground aquifers to solve the primary problem of water supply. The enormous capital expenditure and technical expertise that these schemes would require, however, might be more fruitfully directed to better-watered lands which are undoubtedly capable of greater production than at present. Only in the newly-rich Middle East or parts of India and Pakistan are conditions favourable at present for extensive colonisation of the desert lands, and to see the rest of the world's deserts as a panacea for the world's problems is yet another of the mirages which in popular western mythology are associated with deserts.

References and Further Reading

Most ecology texts contain adequate accounts of plant response to aridity. Some texts which range fairly widely in their examples are listed below.

**Cloudsley-Thompson, J. L. and M. J. Chadwick 1964. *Life in deserts.* London: Foulis. (A very well-illustrated text at an introductory level.)

**Daubenmire, R. F. 1976. *Plants and environment,* 3rd edn. New York: Wiley.

Rodin, L. E. and N. I. Bazilevich 1967. *Production and mineral cycling in terrestrial vegetation.* Edinburgh: Oliver & Boyd.

*Unesco 1960. *Plant-water relationships in arid and semi-arid conditions.* Paris: Unesco.

**Walter, H. 1971. *Ecology of tropical and subtropical vegetation.* Edinburgh: Oliver & Boyd.

Watts, D. 1971. *Principles of biogeography.* London and New York: McGraw-Hill.

Specific aspects

Gates, D., L. A. Stoddart and C. W. Cook 1956. Soil as a factor influencing plant distribution on salt deserts in Utah. *Ecol. Mono.* **26,** 155-75.

Johnston, I. M. 1941. Gypsophily among Mexican desert plants. *J. Arn. Arbor.* **22,** 145-70.

Mitchell, J. E., N. E. West and R. W. Miller 1966. Soil physical properties in relation to plant community patterns in the shadscale zone of north-west Utah. *Ecol.* **47,** 627-30.

10

The temperate deciduous forests

The deciduous forests of the northern hemisphere have been profoundly changed by long-established and repeated clearance for settlement and agriculture, even in relatively recently settled North America, so that only remnants remain from which to piece together their original character. Some fragments, such as the much-studied cove hardwood forests of Tennessee, are richer in species than others, so that conclusions about their ecology may not be strictly comparable from area to area. The climatic conditions under which they flourish can also vary considerably, so that marked differences in growth rates can be recorded as between one area and another. However, in spite of their varied habitats and species composition, their structure is remarkably similar, as can be seen in Figure 50.

One puzzling aspect about these forests is the fact that they occur only in the northern hemisphere. Areas in the southern continents with similar climates are largely dominated by evergreens. It seems possible that this striking difference of biogeography may have its origins in an accident of evolution in the remote past. In the Cretaceous period for some reason—possibly as a response to drought—the angiosperm trees may have developed the deciduous habit during their spread into the northern hemisphere, but not in the southern lands (Axelrod 1966).

10.1 Biomass, production and mineral cycles

Quite a number of studies have now been made of the ecosystem structure and dynamics in these forests and sample figures are set out in Table 4.

In general, complex communities appear to be able to reach higher biomass figures than simple communities. Thus some of the figures obtained in North America indicate a steady biomass increase until the community is 200 years old, when it appears to level off around 400 t/ha^{-1} (Whittaker and Woodwell 1968). On the whole, figures quoted

155

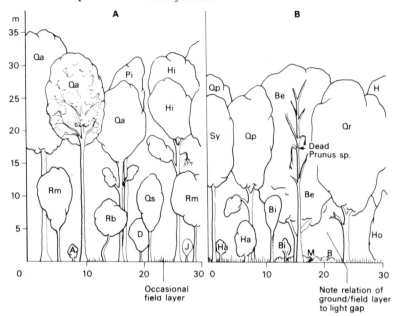

Figure 50. Mid-latitude deciduous summer forest profiles (source: author). **A**—woodland near Roanoke, Va. (leached brown earths over sand and gravels): Qa—white oar (*Quercus alba*); Qs—post oak (*Q. stellata*); Hi—hickory (*Carya glabra*); Rm—red maple (*Acer rubrum*); Rb—redbud (*Cercis canadensis*); D—dogwood (*Cornus florida*); J—Juniper sp.; A—arrow wood; Pi—pine sp. **B**—woodland in Trent Park, Herts, England (leached brown earths over glacial gravels): Qp—sessile oak (*Q.petraea*); Be—beech (*Fagus sylvatica*); H—hornbeam (*Carpinus betulus*); Sy—sycamore (*Acer pseudoplanatus*); Bi—birch (*Betula pendula*); Qr—Quercus pedunculate oak (*Q robur*); Ha—hawthorn (*Cataegus oxycantha*); Ho—holly (*Ilex aquifolium*); M—mellick grass; B—blackberry. Both transects 10 m wide. Example B was part of Enfield Chase until the eighteenth century and the older trees bear signs of pollarding, e.g. the beech shown.

for Western and Central European forests of comparable age seem to be lower than this. The contrast between the net primary production of these forests and that of the wet tropical forests should be noted. As we saw, the latter at maturity seem to have very little, most of the gross production being consumed by respiration. Thus the rest of the food chain dependent on the net primary production has a small biomass total. However, in these mid-latitude forests, even very old communities produce an abundant surplus, much of which is passed annually to the soil to the benefit of the soil flora and fauna. Some of

Table 4 Biomass distribution and production in temperate deciduous forests				
	Oak/pine forest New York	*Cove forest Tennessee*	*Birch forest UK*	*Beech forest Germany*
Age	40-45	150-400	42	120
Biomass (t/ha⁻¹)	97 (trees)	585 (trees)	47.2 (trees)	185 (trees and undergrowth)
Leaves %	4.2	0.6	1.27	1.4
Trunks & branches %	61.4	85.9	71	73
Roots %	34.2	13.5	27	25
Net primary production (t/ha⁻¹/an⁻¹)	10.6 (trees) 1.3 (undergrowth)	18.00 (trees) 0.9 (undergrowth)	1.8 (trees)	6.5 (trees)

the variations in biomass and production in these forests can certainly be ascribed to climatic factors. Cloudy, oceanic types with lower summer temperatures inhibit production rates when compared with continental climates. To an observer accustomed to the British Isles, the rapidity of growth which can be achieved by woody plants of the Eastern USA in the sub-tropical air masses which dominate the summer weather is astonishing.

The amount of minerals stored in the biomass varies according to age and species. At maturity in American forests it can be as much as 7500 kg/ha^{-1} although most other recorded values are less than this; for example, Central European beech forest at 4196 kg/ha^{-1}, lime forest in southern European Russia at 2721-3771 kg/ha^{-1} and British birch forests at 678 kg/ha^{-1}. In percentage terms, the greatest concentrations of nitrogen and the other major mineral nutrients occur in the leaves. Lime leaves, for example, can yield a percentage dry weight of minerals of 7.23 and in the American elm the figure is usually over 9 per cent.

Although most of these abundant minerals are retained by the plant, the minerals remaining in shed leaves are important in supplying soil animal life with a rich nutrient supply, especially of nitrogen. The annual uptake and return of minerals appears to lie between 200 and 500 kg/ha^{-1}, i.e. about half the values for tropical seasonal forests and only a quarter of those in tropical rain forests. Unlike the situation in tropical rain forests, however, many of the minerals are in only loose circulation. Studies on the Hubbard Brook in New Hampshire show a high loss of calcium especially, output in the streams being four times its input from the atmosphere. (On the other hand nitrogen is tightly held, atmospheric input being twice output.) In tropical conditions this would quickly spell disaster, but in temperate lands rates of rock weathering and lowering of the land surface by erosion ensure that nutrient supply is continually replenished from below, the materials being lifted by tree tap roots, soil creep and, above all, by soil animals.

10.2 Soil types

The rich soil fauna of the **brown forest** (or **brown earth**) soil type characteristic of these formations is also responsible for the constant movement downwards of humus in the soil. The constant vertical mixing ensures—on nutrient-rich material, at any rate—that these soils lack marked horizons (Figure 51). The rich litter supplied to the soil is rapidly mineralised, mostly by animals, in two stages. First, the primary decomposers (millipedes, woodlice, beetles and earthworms)

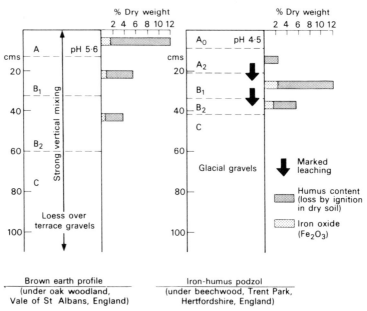

Figure 51. Mid-latitude soil profiles. Note in the brown earth profile the horizons are marked by vague transitions rather than the sharp demarcations observable in the podzol. Also, although the podzol is here found under beech woodland, the profile type is observable in almost the same form under nearby coniferous trees. Brown earth profile: A—dark brown; B_1 and B_2—lighter brown; A to C deeply penetrated by earthworm burrows. Podzol profile: A_0—raw humus; A_2—greyish leached horizon; B_1—dark humus and iron staining; B_2—lighter iron staining (local iron pan nearby); C—no animals present and few roots. (Source: author.)

attack the litter and their faeces are important in producing a good crumb structure as well as providing the food for the secondary decomposers (mites, springtails, and potworms) which complete the process. Wet material is also broken down by bacteria, fungi and protozoa, and the general attack by the soil flora and fauna is so effective that almost complete mineralisation of mild humus is achieved annually, so there is little accumulation on the surface (Figure 45). Only in very wet conditions or with acid humus does litter decompose so slowly, by wetting and bacterial and protozoan feeding, as to produce a peaty surface layer. Profiles in this case are usually marked and the soils may be described as leached brown earths or, with more intense leaching, podzolised brown earths.

Not all soils below these forests are brown earths, however. In the warmer climates of the south-eastern USA large parts of the mixed pine and hardwood forests are associated with red and yellow podzolic profiles. In these soils the prevailing higher temperatures encourage mobilisation of some of the clay fraction of the silicate minerals, and this is leached downwards. The less stable mineral structure in these types makes them more liable than the brown earths to erosion following deforestation.

On a local scale, the drainage characteristics of soil parent material and its ability to release nutrients have some influence on the geographical distribution of species. In Britain, for example, freely drained gravels may be associated with the sessile oak (*Quercus petraea*), beech (*Fagus sylvatica*), sweet chestnut (*Castanea sativa*) and hornbeam (*Carpinus betula*), whilst loams favour the pedunculate oak (*Quercus robur*), the elm (*Ulmus* spp.) and the lime (*Tilia* spp.). On limestones in northern Britain, the ash (*Fraxinus excelsior*) is often dominant and the wettest soils in all areas usually carry alder and willows. However, there are no hard-and-fast rules governing any of these distributions. Most of the European species are survivors of very hard times during the Quaternary glacial ages and, as might be expected, are very adaptable.

10.3 Community patterns

The vastly greater number of species in the American and East Asian formations than in the European ensure that community associations are much more complex in these forests. Within the American formations, three major climax types have been recognised:

(a) In the north, beach (*Fagus grandifolia*) and maple (*Acer saccharum*) dominate with admixtures of conifers like hemlock;

(b) To the south-east, oak (*Quercus montana, alba*, etc.) and chestnut (*Castanea dentata*) dominated forests are frequent;

(c) To the south-west and west around the southern Appalachians and extending towards the prairies, oaks (*Quercus borealis, velutina, alba, macrocarpa* and others) and hickories (*Carya* spp.) are the most conspicuous elements in the forests. There is some evidence that this zonation may be the result of the response of plants to both the northwards temperature gradient and the east-west moisture gradient. Within the forests most associations can be broadly correlated with particular moisture and evapo-transpiration values (Figure 39).

The Asian formations are composed of the same genera (apart from hickory), particularly important species being the native ash (*Fraxinus mandshurica*), birch (*Betula armanii*) and beech (*Fagus crenata*).

In Europe, the common forest species of the north-west are replaced towards the Mediterranean by other deciduous oaks such as the pubescent oaks (*Quercus pubescens*) accompanied by evergreen oaks. Towards the east, the beech is absent beyond the Vistula and from the Ukraine to the Urals, the rich loess soils are dominated by the oak (*Quercus robur*), the sandy soils by Scots pines (*Pinus sylvestris*). At the northern limit of the forests, the dominant tree in southern Scandinavia is the beech.

10.4 Man and the deciduous forests

Since the introduction of agriculture to these forests by the Neolithic farmers (about 4800 BC in Britain (Smith 1974)) the soils and forest resources of this formation-type have been profoundly transformed by the activities of Man, but nowhere has the forest disappeared completely. In Europe and the Far East, for many centuries, the forests were incorporated into human farming and land management practices adapted to the sustained conservative use of the forest resource. The economic system was controlled by an armoury of checks and legal rules which attempted to combine the good of the commonality of people and forest with the needs of the individual for fuel, building timber and pasturage. Systems such as these still operate in parts of Europe and possibly China, but in lowland England and Denmark, for example, many of them were swept away at the time of land enclosure and dispersal of the farms from nucleated villages. Tenuous fragments of these systems survive, as in the New Forest, but even here the system has been disrupted in response to commercial pressures (Tubbs 1970). In North America, the settlement of the eastern forests never produced the extensive clearances of Europe and China, and since the 1940s there has been, in any case, a reversal of the process. The increasing efficiency of mid-western farming has resulted in the abandonment of much of the eastern farmland to tax-loss investment or urban development (Hart 1968) with the rapid re-invasion of trees of the early seral stages (Figure 52). Probably there has never been so much land covered by the deciduous forests in North America since the beginning of the century.

The soils of this formation are superb providers, responding to good husbandry with generous crop yields. These yields are not simply the

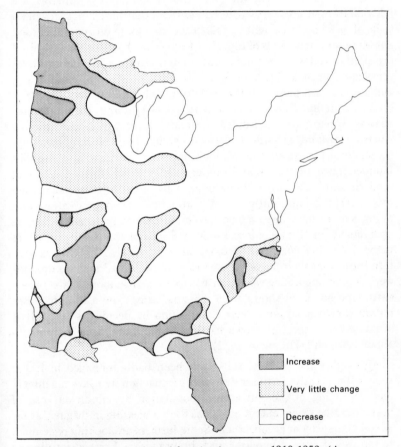

Increase

Very little change

Decrease

Figure 52. Change in cleared farmland acreage, 1910-1959. (Areas generalised from Hart (1968). Copyright Association of American Geographers.) The factors underlying change on this map are very complex and include increasing urbanisation and industrialisation and the building of roads as well as the abandonment of marginal farming land in the poorer soil areas in the east under competition from more efficient farming in the Mid-West.

product of ingenious technology or the addition of chemical fertilisers. Their natural drainage characteristics, well-aerated structure, and abundant microbial and animal life give the soils tremendous strength even when abused by heavy machinery and excessive chemical treatment. No better example could be found than the famous Broadbalk field at the Rothamsted experimental station (at Harpenden, England) which has produced cereal crops for over 100 years without the addition of fertiliser of any kind. Its nitrogen supply is maintained by the blue-green algae coating the soil crumbs, and other minerals are released from its great reserve of bases. In all that time it has shown no signs of eroding and might appear to be able to go on supplying food almost indefinitely. Soils of this kind formed the base for the growth of prosperous states in northern and western Europe in late medieval times and European history has revolved around the centres of wealth and power which these soils fostered down to the present day. The Paris Basin, Aquitaine, Burgundy, Flanders, eastern Jutland, the Rhineland, Saxony, Westphalia, the English lowlands, Bavaria, Bohemia, and the Austrian lowlands are still regions with political and cultural meaning, especially as the growth (almost literally) of wealth based on these soils founded most of the great European cities which focus the life of modern man. The less well-endowed soils, especially the podzolised brown earths on nutrient-poor substrates, are more open to abuse and at times in the past have even been so impoverished by farming as to be abandoned. In Britain, for example, the soils of the North Yorkshire moors overlying coarse sandstones were overgrazed by Bronze Age farming so that their nutrient reserves were exhausted and the uplands were depopulated. The raising of their status once more to the brown earth type has been shown to be possible using only a little fertiliser initially and encouraging their colonisation by deciduous trees. It is estimated that the full brown earth ecosystem could be restored in about 100 years (Dimbleby 1952).

Although some of these soils have been badly managed in recent years with excessive fertiliser dressing, compaction by heavy machinery and degrading of the soil animal population by injudicious use of pesticides, the recent energy crisis has been a blessing in disguise as far as soil management is concerned. Some farmers are looking once more to the use of animal manures and the practices of mixed farming to redress some of the imbalance encouraged by freely available cheap energy, and these changes can only be to the benefit of the soil ecosystem.

Probably a greater threat to the soil is the demand for land for urban development and roads. It is probably true to say that far too little consideration is given to soil resources when making planning decisions about the designation of land use. Thus, in Britain the second Land Utilisation Survey has revealed how much of the land—even with higher housing densities than during the 1930s—has been wasted by poor planning (Coleman 1977). Whether in future, western societies will give greater consideration to soil resources and treat them with some of the respect shown by our medieval forebears is dependent on soil being seen as the fundamental resource on which our human ecosystem depends, and which underpins the whole of the rest of our activities. The prospect of the world's population doubling in the next thirty years should, as Dr Johnson said of the prospect of hanging, focus the mind wonderfully. Given adequate understanding of what nature has done for us, there is no real reason why the forests and soils of this formation-type should not continue to provide Man with their economic, recreational and aesthetic benefits as they have done so effectively for 6000 years.

References and Further Reading

General

*Eyre, S. R. 1968. *Vegetation and soils.* London: Edward Arnold. (A full account of the vegetational character of these formations.)

Perring, F. (ed.) 1971. *The flora of changing Britain.* London: Classey. (Useful collection of case studies.)

**Tansley, A. G. 1968. *Britain's green mantle, past present and future,* 2nd edn (revised by M. C. F. Proctor). London: George Allen & Unwin. (In its revised form still the best introduction to the vegetation of the British Isles. A masterly work.)

Tubbs, E. C. 1970. *The New Forest; an ecological history.* Newton Abbot: David and Charles. (A detailed case study of a small region illustrating the complexity of the interrelationships between Man and vegetation in these small crowded islands.)

Specific aspects

Coleman, A. 1977. Land use planning: success or failure? *Architect's Journal* **165**(3), 94-134.

Dimbleby, C. W. 1952. Soil regeneration on the north-east Yorkshire moors. *J. Ecol.* **40**, 331.

Hart, F. J. 1968. Loss and abandonment of cleared farmland in the eastern U.S.A. *Ann. Assoc. Am. Geogs* **58, 417-40. (A clear exposition of the many factors which result in the loss of farmland.)

*Mather, J. R. and G. A. Yoshioka 1968. The role of climate in the distribution of vegetation. *Ann. Assoc. Am. Geogs* **58,** 29-41. (An interesting attempt to relate PE and other factors to vegetation.)

Ovington, J. D. 1962. Quantitative ecology and the woodland ecosystem concept. In Craggs, J. B. (ed.) *Advances in ecological research.* New York: Academic Press.

Smith, I. F. 1974. The neolithic. In Renfrew, C. (ed.) *British prehistory.* London: Duckworth.

Whittaker, R. H. and G. M. Woodwell 1968. Dimension and production relations of trees and shrubs in the Brookhaven Forest, New York. *J. Ecol.* **56,** 1-25.

Whittaker, R. H. and G. M. Woodwell 1969. Structure, production and diversity of the oak-pine forest at Brookhaven, New York. *J. Ecol.* **57,** 157-76.

Texts dealing with the ecology of lands once forested but now kept in some form of broken climax are numerous. Two especially recommended are:

*Gimingham, C. H. 1972. *The ecology of heathlands.* London: Chapman and Hall.

*Spedding, C. R. W. 1971. *Grassland ecology.* London: Oxford U. Press.

11

The coniferous forests of the northern hemisphere

The great tracts of boreal forests of the cold temperate lands are not the only lands dominated by conifers. As Figure 57 shows, on the western seaboard of North America, lands with cool, very damp climates have extensive coniferous forests. In mountainous regions, the sub-alpine stages are dominated by conifers. There are extensive coniferous forests in Brazil and they are important in the Mediterranean basin. Clearly all these types are not of the same kind, nor are they responding to the same environmental circumstances. In one formation it may be their resistance to drought which is important; in another, their resistance to low temperatures; in yet another, their resistance to fire. The characteristics are set out in Table 5, but it is with the best-known type, the boreal forests, that this section is most concerned.

11.1 Biomass, productivity and mineral cycles

The mean biomass of mature boreal coniferous forest seems to be around 100t/ha^{-1}, although the range is wide, from 25-175t/ha^{-1}. The leaf portion is greater than that in the deciduous forests, and in boggy areas and low-quality open forest the contribution of the ground flora to the green biomass can be very large indeed, reaching 40 per cent in some wooded swamps with considerable bog moss (*Sphagnum*) growth. There is some confusion as to just how great biomass can be. Rodin and Bazilevich (1967) cite well-managed plantations of larch about 140 years old as having a total plant biomass of 800t/ha^{-1}. Net primary production varies greatly according to age and quality of stand, species composition, climatic conditions, ground conditions and the degree of waterlogging. The range seems to be from 4-20t/ha^{-1}/an^{-1} with a mean around 8t/ha^{-1}/an^{-1}. Thus, spruce usually displays higher net primary production than pine, but this cannot be solely ascribed to the differential capacity of the two as primary producers. As can be seen from Table 5, pines and spruce prefer different soil types, the spruce

165

Table 5 Major coniferous Formation-types

Type	Location	Botanical character	Environmental relationships
Boreal forest	N. America and Eurasia	Species of pine, spruce, fir and larch plus birch, alder and poplar	i) Vegetative life confined to four months approximately ii) Strongly xeromorphic to withstand physiological drought of winter iii) Deciduous larches (e.g. the Dahurian larch of Siberia) are often associated with the most rigorous conditions iv) Local edaphic climaxes occur, for example, white spruce (*Picea glauca*), balsam fir (*Abies balsama*) and European spruce are associated with deeper loams, the jackpine (*P. banksiana*) and Scots pine (*P. sylvestris*) with poorer sandy types.
Sub-alpine types	Himalayas, Alps, Caucasus, Rockies, Carpathians and Cascades	The genera and in some cases the species of the boreal forest	The growth season and the length and climatic severity of winter are broadly comparable to those in the boreal forest.
Western N. American coastal forest	From Alaska to California	Many species and sub-species with Sitka spruce (*P. sitchensis*) often dominant from Alaska to British Columbia, redwoods in Oregon and N. California and Douglas fir (*Pseudotsuga taxifolia*) in British Columbia and Washington	With such a span of latitude, the main environmental influences vary considerably. The distribution of redwoods, for example, can be closely correlated with fog incidence whilst the Douglas fir is associated with the liability to fire.
Montane forest of N. America	Interposed between coast forest and sub-alpine types	Ponderosa pine (*P. ponderosa*), Douglas fir, white fir (*Abies concolor*) and redwood (*Sequoiadendron sempervirens*) are the main dominants	One of the most important ecological relationships is to fire—most of the species are fire-resistant. In most of the climatic zones strong summer drought favours the coniferous habit.
Lake forest of N. America	Minnesota to northern New England	Originally dominated by white pine (*P. strobus*), red pine (*P. resinosa*) and eastern hemlock (*Tsuga canadensis*)	Only secondary forest remains now, so environmental relations are obscure.
Pine barren of S.E. USA	N. Florida, Alabama, Texas	Loblolly pine (*P. taeda*), shortleaf pine (*P. Echinata*), pitch pine (*P. rigida*), longleaf pine (*P. palestris*) and slash pine (*P. caribenea*) dominate	Forests are doubtfully a climatic climax. They may be, at least in part, fire climax maintained by aboriginal inhabitants.
'Mediterranean types'	Mediterranean basin, California, S. Chile	Cedars and pines in Mediterranean basin, pines in California, Araucaria pines in Chile	All are associated with strong summer drought. In the Mediterranean basin some of the lowland pines represent a stage towards angiosperm forests.

usually occupying the more eutrophic soils such as deep loams over glacial clays, which encourage a greater production than the more oligotrophic sites occupied by pines. Moreover, experiments in Sweden to test the response of forest trees to the effects of fertilisers showed significantly higher levels of production in Scots pine than those usual in natural conditions. The cultivation of lupins within the pine stands, for example, more than doubled net primary production in fifty-year-old trees, an improvement which was sustained for at least ten years (Stålfelt 1972).

The nutrient cycles of the boreal forests are significantly different from those of the deciduous forests. In the first place, the needles of the trees—apart from larch—are much longer-lived than the leaves of deciduous trees, and over their life they lose nitrogen and other elements by leaching. Thus, the litter returned to the forest floor has few elements that would neutralise the products of humification, and an acid humus results. The needles also carry waxes and resins which are not readily eaten by animals or easily humified. In any case, the generally low temperatures, short active season, often poorly-drained ground and low energy content of the litter inhibit the development of a flourishing soil fauna. Thus, one of the outstanding features of coniferous forest ecology is the mass of undecomposed litter on the soil surface. In the more northerly regions it may be three to five times the annual litter fall, and in swamp forest it may exceed the annual litter fall by twenty to thirty times. The mineral element circulated in the greatest quantity appears to be calcium. (See Figure 53 for pattern of circulation of this element in pine forest.)

11.2 Soil conditions

The release of acids from the litter and its poor humification and mineralisation rates ensure that the formation of a stable colloidal complex with good reserves of nutrients is almost impossible. The acids migrating in the profile react with cations in the soil minerals to form complex compounds which are leached downwards to accumulate in what is called the humus-illuvial or iron-humus-illuvial horizons (Figure 51). As there is no significant upwards mixing in the soil there is an irreversible loss of nutrients which is not fully compensated for by atmospheric input or migrating birds and other animals. Given this state it is interesting to speculate whether the soil can run out of nutrients altogether. There seems to be evidence that this can happen in certain circumstances. Thus, experiments in Sweden show that in some pine heaths, seedling growth is dependent on nutrient supply and that mobile reserves in the soil are sufficient for *only one* generation.

Note: pool and transfer values drawn to same scale

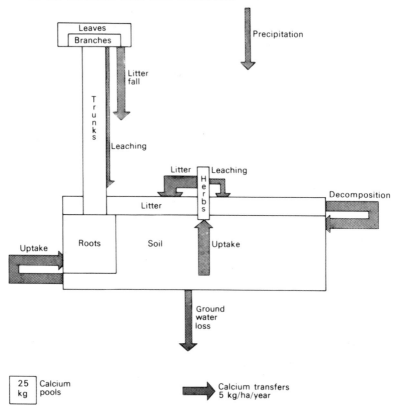

Figure 53. Calcium pool and transfer values for mature, stable Scots pine forest, Gt. Britain (data from Ovington (1962)). Note that precipitation input and ground water loss are practically equivalent.

Regeneration is only possible if this reserve is supplied by litter fall from the existing trees. If these are cut down the reserve is lost with them and regeneration ceases (Stålfelt 1972). This is an extreme case, of course. In most forests, available reserves in the soil or from ash are usually sufficient to ensure regeneration after felling or burning.

A vital role in maintaining the nutrient cycles in these forests is played by the mycorrhizal fungi. They infest the litter and ensure that the mineral elements are transferred directly to the host tree before the nutrients are leached downwards out of reach. In fact, there are many similarities in their role in these forests to the part they play in the hot wet forests. There is good evidence that mycorrhizae represent a major pathway from litter to tree for nitrogen.

11.3 'Mixed' forests

The transition from coniferous to deciduous forests is very wide, so wide in fact that many authorities recognise it as a distinct formation-type, often called **ecotone mixed forests.** However, the conifers and deciduous trees do not form mixed stands readily. Instead, a mosaic pattern of distinct association occurs, dominated by one or the other vegetation-type. Thus, in European Russia from 59° southwards, on the deeper loams the spruce is replaced by the oak (*Quercus robur*), but the poorer sandy soils continue to be dominated by pines as far south as the Ukraine. Except on a very broad scale, the continuum concept of vegetation is difficult to recognise in these forests. Their pattern is much nearer the polyclimax envisaged by Tansley. A similar pattern of fairly discrete associations can be recognised in North America where sandy outwash lands in Minnesota, Wisconsin and Michigan were originally dominated by red and white pines, whilst on the deeper loams oak and hickory associations flourished. In the Far East, in northern and eastern Manchuria, northern Korea and northern Japan, the flora is richer than in Europe or North America and the mosaic more complex with many more truly mixed stands.

The boreal forests represent one of the world's great ecosystems into which Man has ventured in large numbers only recently, and for the most part they are still wilderness. Only in Europe and here and there in North America and the USSR have extensive inroads been made. As these formations lie almost wholly within the purview of advanced industrial societies and a great deal of knowledge has already been accumulated about their management for continuous timber production, their future should be assured. It is probable that the enormous cost of improving their soils for agriculture to counter the impoverished nutrient cycles is likely to prevent the development of much farming, except perhaps under extreme population pressure in China and Japan. Besides which, in North America and Russia, there are still extensive lands with better soils which have not yet reached their full agricultural potential. (See, however, Azhiev (1976) for an account of damage to the Siberian taiga.)

The mixed forests, on the other hand, have been extensively cleared already in eastern Europe and around the Great Lakes and their brown earth soils especially have been made almost as productive as those in western Europe.

References and Further Reading

Many texts on the Earth's vegetation carry full accounts of coniferous forest types. Particularly recommended is:

**Stålfelt, M. C. 1972. *Stålfelt's plant ecology*. London: Longman.

Also see:

FAO *Handbook on forest inventory*. Paris: Unesco.

Sukachev, V. and N. Dylis 1964. *Fundamentals of forest biogeocoenology* (English trans. J. M. Maclennan). Edinburgh: Oliver & Boyd. (This text contains a mine of data on both coniferous and deciduous forests.)

Whitmore, T. C. 1975. *Tropical rain forest of the Far East*. London: Oxford U. Press. (Good account of conifers in the tropics, a subject dealt with in few books.)

Specific aspects

Azhiev, M. E. 1976. Taming the Siberian taiga. *New Sci.* **69**(988), 382-5.

Clausen, J. 1965. Population studies of alpine and sub-alpine races of conifers and willows in the Californian High Sierra Nevada. *Evolution* **19**(1), 55-68.

Rodin, L. E. and N. I. Bazilevich 1967. *Production and mineral cycling in terrestrial vegetation* (English trans. G. E. Fogg). Edinburgh: Oliver & Boyd.

12

The temperate grasslands

The majority of the lands formerly covered by the temperate grasslands are now under cultivation or controlled grazing. Only the drier edges are comparatively little affected. It is true that there are reserves in North America, the USSR and Central Asia, some of considerable size, where the ecosystem comes close to its original nature, but even here the animal populations integral to the ecosystem have been much reduced. Consequently, it might be thought that the descriptions of the ecology of these lands are only of academic interest. However, in one important respect this is not so. Many of these lands have suffered in the past precisely because of the lack of this 'academic' knowledge. Even if the species which composed the living part of the ecosystem have gone for ever, their shadow is still cast in the soil. Moreover, not only is ecological knowledge essential for the maintenance of the soil in good heart; the essential preservation of the genetic resources of the plant and animal species forms a reservoir for breeding, and a number of centres have been established for this purpose in the USA (Watt 1968).

12.1 Vegetational features

Physiognomically, the steppes or prairies are usually dominated by more or less xeromorphic grasses with intensive rooting systems mixed with other life-forms including herbaceous annuals and perennials, bulb plants and woody shrubs. They occur under a variety of climatic regimes in mid-latitudes, all characterised by a dry season. Formations to which the term 'steppe' has been applied have also been described as occupying the Sahel zone at the southern Sahara fringe, parts of Namibia and south-west Australia. The term 'pseudo-steppe' has also been used for such formations.

The particular structure and floristic character of these grasslands vary according to climate, soil type and biotic factors, but it must be noted that the ecological status is very difficult to determine exactly in many cases. There is good evidence from many parts of the world that fire has had a considerable part to play in their maintenance and that much of the firing was induced by Man.

Also, it must be emphasised that the development of any grassland is not a simple response to climate or any other single factor. Their existence depends on a 'complex of selective forces' (McMillan 1959), no small part of which was the enormous population of grazing animals which formed an integral part of the ecosystem. Moreover, the differentiation in feeding niche amongst these animals ensured that there would be few herbaceous or woody species unaffected by grazing or browsing. The populations carried by these formations, even such a comparatively short time after their demise, are now difficult to imagine. The buffalo population has been estimated at between 50 and 60 million head before its decimation, each of them consuming almost as much as a cow. These animals shared the land with large numbers of pronghorn antelope and others such as the rodent 'prairie dogs' (*Gnomys*), which were organised into societies with territories covering several square kilometres. They consumed enormous amounts of plant material; for example, 87 per cent of the total production of a steppe area studied in Arizona. It would have been a very resistant woody plant indeed which could have reached any great height in face of the irresistible combination of drought and feeding pressures.

The dominant grasses of the northern mid-latitude steppes belong to the genera *Stipa, Agropyron, Koeleria, Festuca, Andropogon* and *Elymus* and these, with others such as *Themeda* in the South Africa grassveld and *Melic* in the Pampas, also occur in the southern hemisphere. The better-watered steppes with rainfalls around 2000 mm/an can carry taller grasses with deeply penetrating and highly ramified roots. In the American tall-grass prairies, the common dominants such as *Andropogon* spp. and *Stipa spartea* reach two metres or so. With increasing drought along the moisture gradient towards sub-arid regions the tall-grass steppe gives way to a mixed steppe of tall and low forms and finally to short-grass formations where the dominant Buffalo grass (*Buchloe dactyloides*) and grama (*Bouteloua gracilis*) reach only a few inches. A similar gradient was originally observable in the USSR.

Reference has already been made (Chapter 2.4) to the ecotypical zonation into long and short-day types of these grasses which occurs with latitude, and studies show that the apparent uniformity of the grasslands conceals the fact that they are a mosaic of ecotypic variant species. McMillan (1959) has pointed out that other ecotypic variations, for example, in relation to soil, lead to the grasslands displaying 'a geographic repetition of certain species and genera over an obvious non-uniform habitat'.

The southern hemisphere varieties of steppe land occur under a range of climatic regimes and are all of doubtful status as climatic climax formations. The pampas type of southern Brazil, Uruguay and Argentina, which originally displayed something of the division into long and short-grass types, has almost entirely disappeared as the original bunch grasses (*Melic* spp.) have been replaced by European turf-formers. In areas like Uruguay and north-east Argentina with apparently more than enough rainfall to support trees, they may, in any case, have been a fire climax maintained by the original Amerindians. Similar suggestions have been put forward less convincingly for the South African grassveld and the Canterbury Plains of New Zealand. In Australia, although there were originally wide areas of grassland composed of mixed tropical and mid-latitude species, these are often continuous with the tree-savanna and the grass and shrub strata of open Eucalypt woodlands, so they resemble those of the northern hemisphere only superficially.

12.2 Biomass, production and mineral cycling

Total plant biomass lies in the range 20-50t/ha^{-1} for what Russian authors describe as 'plakor steppe', i.e. flat or slightly undulating, well-drained grassland above valleys and ravines. Although considerable variation occurs from year to year according to wetness of the season, the total plant biomass remains constant overall. This is due to the carry-over from one season to another of root material whose decay is greatly reduced in dry years. Total net primary production varies between about 1.5 and 15t/ha^{-1}/an^{-1} with a mean around 5t/ha^{-1}/an^{-1}. As in the savanna, production is closely related to rainfall, but within quite short distances the vegetation mosaic produces surprisingly high variations. In North Dakota, for example, the tall-grass type has three times the production of short-grass association nearby (Hadley and Burcos 1967). Data from the USA and USSR show that biomass varies very little from one steppe type to another. Along the moisture gradient from wet to dry lands the falling proportion of above-ground material is balanced by the increase of root material.

In an ecosystem dominated by herbs, the litter amounts are naturally very great, usually over 40 per cent of the total above and below ground organic matter. The amounts of nutrients—especially silicon, nitrogen, calcium, potassium and phosphorus—returned to the soil are considerable. In fact, the long-grass prairie (steppe meadow) returns the largest amounts of nutrients to the soil annually of all the temperate vegetation types. The rate of decomposition is usually somewhat slower than its supply, so most soils have a thick sod or 'steppe matting' several centimetres thick whose mass is usually larger than the total mass of the above-ground green parts. The amount of matting varies from season to season but always exerts a considerable effect on the soil moisture balance. It interferes with percolation and intercepts and holds water which falls onto it. As evaporation rates are highest in the rainy season (summer) little downwards percolation takes place and much of the water which finds its way into the soil is from melting snow. Consequently, the sub-soils of the steppes tend to be very dry, some of them almost permanently so. In moister regions snow melt and heavy summer storms do produce downwards movement of the most mobile soil compounds, for example, sodium and potassium salts. In the **prairie soil** type of the USA the amount of leaching is sufficient to produce a slightly acid reaction.

Steppe matting is also effective in preventing erosion. Water wash is prevented and so is wind blow, as the fine, often silty particles characteristic of the soils are never in direct contact with the atmosphere. Where matting is removed by ploughing or overgrazing, its protective function has to be compensated for in some way; for

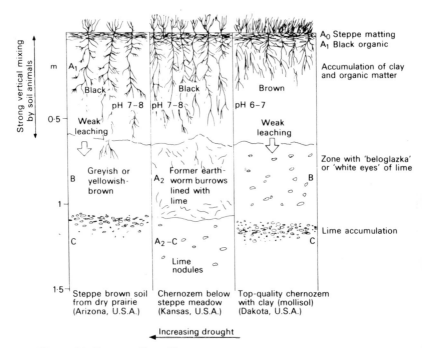

Figure 54. Steppe soil profiles.

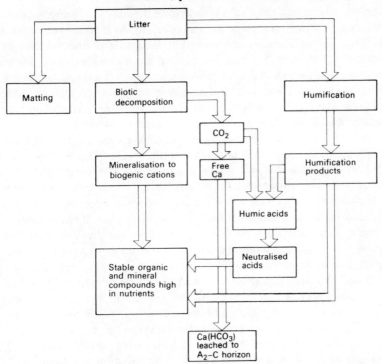

Figure 55. The process of litter decomposition in chernozem soils.

example by mulching, i.e. covering the soil with straw and leaves, or by wind-break planting as in the southern Ukraine, or by contour ploughing or combinations of techniques suited to local needs.

12.3 Soil types

Three widespread major soil types are associated with the temperate grasslands of the northern hemisphere, and their main features are set out diagrammatically in Figure 54. The processes which result in one of the most characteristic features of these soils, the accumulation of lime, are explained in Figure 55.

When men first broke these soils with their ploughs, the initial high yields were sometimes followed by widespread soil erosion by wind and water. In the USA in the 1920s, in Australia in the 1940s and in Kazakhstan in the early 1960s, disastrous losses of soil occurred because land management practices ignored the realities of ecology. In America, studies by ecologists, agronomists, hydrologists and engineers have gradually built up adequate measures for conserving the soil in good heart, and these are embodied in the work of the Federal US Soil Conservation service backed by the power of Federal law. By the early 1950s over 75 per cent of the arable lands of the United States were covered by strict Federal conservation laws. The tragic loss of soil from the plains of Soviet Central Asia in recent years occurred precisely because the lessons learned at such painful cost in terms of human misery and dollars in the United States were not heeded.

Although the great herds of these plains have gone for good, the soil populations are still there, and so in the drier edges of these lands are the plant species. As part of the management of these lands in the USA, breeding programmes are encouraged to improve the range-plants particularly (Watt 1968). These programmes represent a model for many ecosystems in the future. The fact that the United States has turned what, only a few decades ago, were virtual disaster areas into some of the most productive land in the world shows what can be done, given political will allied to scientific resource management.

That such practices should incorporate, as an integral part of their activities, the conservation of genetic material is an increasingly urgent need in all parts of the world. In the 1920s the great Russian botanist N. I. Vavilov mapped the homelands of many of our major crop plants and traced their lineage to their wild ancestors (see Vavilov 1952). Recent studies show that many of these wild ancestral types are rapidly disappearing. Once lost, their genes are lost for good and the highly bred food plants cannot be back-bred with the wild types which may contain just the genes required to cope with environmental hazards like disease or low rainfall. The proposal to establish plant gene banks under the aegis of the FAO has been enthusiastically received, and the model for such banks has been established in Italy at Bari. In the next decade or so the needs of the world for a network of such stations will pay handsome dividends on the modest investment involved (Heslop-Harrison 1974).

12.4 The grassland-forest ecotone

The junction of grassland and forest in the temperate latitudes is one of the most complex and interesting transitions of any amongst the world's vegetation-types. It is also one of the most puzzling. From some accounts it was possible in places to walk from closed forest to open grassland in the space of a few yards. In other places, islands of forest existed out in the open prairies and the patches of grassland amongst the forest created a park-like zone tens of miles deep.

Many workers have advanced hypotheses to explain the intricate nature of this ecotone and most are agreed that a simple climatic explanation is unsatisfactory. Fire and human interference have undoubtedly had some influence, but these are insufficient by themselves to account for the characteristics of the transition. In the Canadian park-zone, for example, the aspen is fire-tolerant and can reproduce by suckers. Also, in Alberta, the Ukraine, the Danube basin and European Russia, relict chernozem soils (called **grey-brown forest soils**) exist under closed forest, indicating invasion of former grasslands by trees. Why this happened is impossible to say. In America it certainly took place before the European settlement. Extreme views as to the anthropogenic origin of grasslands are hard to sustain in face of the facts that: (a) steppes have been features of the world's vegetation in epochs other than the present inter-glacial (Figure 7), and (b) the evidence that the community structures of the northern hemisphere grasslands are so complex and well-integrated that they are hardly likely to have come together by chance. In the view of some ecologists (Walter 1971, see p. 47), trees of the temperate lands—as successors to the Arcto-Tertiary forests—are all of closed forests and cannot possibly maintain themselves in competition with grasslands without the preceding successional stages to closed forest. In this view it is these preceding stages which are likely to be least successful in competition with grasses. Yet in Canada, the aspen succeeds in establishing itself very well. Perhaps as these lands become further incorporated into farming systems this problem is destined to become one of the great unsolved puzzles of biogeography.

References and Further Reading

The classic account of temperate grassland is:

**Weaver, J. E. and F. E. Clements 1938. *Plant ecology,* 2nd edn. New York and London: Macmillan.

See also:

Watt, K. E. F. 1968. *Ecology and resource management.* New York: McGraw-Hill.

Vavilov, N. I. 1952. *The origin, variation and immunity and breeding of cultivated plants* (trans. K. S. Chester). Waltham, Mass.: Chronica botanica.

Specific aspects

*Daubenmire, R. F. 1968. The ecology of fire in grasslands. In Craggs, J. B. (ed.) *Advances in ecological research*. New York: Academic Press.

Hadley, E. B. and R. P. Burcos 1967. Plant community composition and net primary production within a native East Dakota prairie. *Am. Midland Nat.* 77(1), 116-27.

*Heslop-Harrison, J. 1974. Genetic resource conservation: the aims and the means. *Roy. Soc. Arts J.* **CXXII,** 157-69.

McMillan, C. 1959. The role of ecotypic variation in the distribution of the central grasslands of North America. *Ecol. Mono.* **29, 285-308.

*Ovington, J. D. 1963. Plant biomass and productivity of prairie, savanna, oakwood and maizefield ecosystems in central Minnesota. *Ecol.* **44,** 52-63.

13

High latitude and high altitude vegetation

13.1 Vegetation types

Arctic tundra and high alpine vegetation are complex mosaics of very different communities, each fragment being related to the local microclimate and to soil type. Arctic tundra is more varied than high altitude tundra with many communities intergrading in an intricate fashion, and some of these are set out in Table 6. The proportion of ground covered by each type varies latitudinally between low, mid and high Arctic, but everywhere it is the influence of local habitat variations which is paramount in determining the community patterns. The ecotone from forest to tundra is rarely abrupt and outliers of forest trees may occur on well-drained sites to considerable distances beyond the forest edge—stunted white or black spruce (Tamarack) in Canada, Dahurian larch in Siberia and the low birch, *Betula odorata*, in Scandinavia.

In contrast with the Arctic, the islands surrounding the Antarctic continent are treeless to much lower latitudes. Islands far distant from the continent may carry a vegetation whose physiognomy closely resembles tundra, although the two flora bear little relation to each other. Their resemblance is an excellent example of convergent evolution under extreme habitat conditions (see Chapter 4.6(1)).

In middle latitudes, the alpine stages on high mountains are comparable in many ways, floristically and physiognomically, to the vegetation near sea level in the low Arctic and there are many common species represented as latitudinal ecotypes (Figure 21). However, important differences in energy conditions and other factors at work in the two ecosystems make direct comparison impossible. For example, primary production in alpine sites is probably less than half that of the low Arctic tundra. Moreover, data for energy and moisture balance, ultra-violet and cosmic radiation indicate that these environments 'undergo one of the greatest energy fluctuations observed on this planet' (Terjung *et al.* 1969). However, the large number of forms

179

Table 6 Major Arctic and alpine physiognomic types

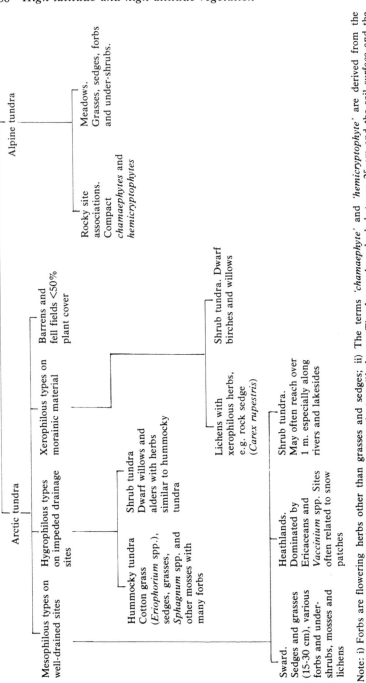

Note: i) Forbs are flowering herbs other than grasses and sedges; ii) The terms *'chamaephyte'* and *'hemicryptophyte'* are derived from the classification by the Danish botanist Raunkiaer of plant life-forms. The former have buds between 25 cm and the soil surface and the latter have buds at the soil surface; iii) In the Arctic on rocky sites near permanent ice caps there are many lichen-free areas now identified ~~as those of the 'Little Ice Age' of the sixteenth and seventeenth centuries (Barry *et al. Science* 190:979).~~

peculiar to the alpine stage means that this formation is richer floristically than the Arctic tundra.

Low-latitude mountains may also have an alpine stage which in some cases resembles tundra, being composed of low herbaceous perennials, and Arctic genera may be present, but there are few Arctic species. The low-latitude high mountains have in addition vegetation-types not present in the Arctic or mid-latitudes for example, the sentinel-like *Senecios* and *Lobelias* of the African volcanoes and the tall grasslands with woody shrubs of many well-watered tropical mountains.

13.2 Biomass, production and mineral cycles

The range of plant biomass in Arctic tundra is very wide; for example, shrub tundra measured at $63°N$ $168°E$ gave a figure of $14.3t/ha^{-1}$, but tundra with similar physiognomy measured at $75°N$ $140°E$ yielded only $3.4t/ha^{-1}$. As in desert communities, life is maintained more easily underground and proportions of underground organs are usually 80 per cent or more of the flowering plant biomass. Most communities contain significant proportions of non-flowering plants (mosses and lichens) and in the first example above about 8-9 per cent of the plant biomass comprises non-flowering plants, in the second example about 18 per cent. As might be expected, the totals of dead material are usually large (Figure 49) both above and below ground, often in waterlogged conditions comprising more than 70 per cent of the total organic matter. There are few figures for production in these ecosystem-types. Figures from the low Arctic of Canada, north-eastern Siberia and Sweden range between 250 and 1800 $kg/ha^{-1}/an^{-1}$. Alpine tundras of temperate latitudes yield sample figures usually between 100 and 1000 $kg/ha^{-1}/an^{-1}$.

The mineral cycles in high latitude tundra are unique in being dominated by nitrogen, and this element alone almost outweighs the total of all the other elements cycled in both shrub and herbaceous tundra.

13.3 Tundra soils

The soils of these lands reflect the almost universal waterlogging, short 'warm' periods, reserves of cold, infrequent aeration and low rates of weathering and mineralisation in the frequent **gleying** which occurs and the mass of peat which is usual. A gley horizon is typically blue-grey or grey with rusty brown streaks and mottling, the latter produced by poor aeration and the conversion of ferrous salts to some form of ferric oxide. In grassy and flowering tundra with relatively large amounts of litter fall, greater amounts of cations are supplied to the soil than in shrub tundra. This, plus lesser humidity and better aeration usual in such communities, ensures higher rates of decomposition, and these soils carry a higher soil animal population, even including earthworms. In practically all the main tundra types, alpine and Arctic, a major role in soil restructuring and mineral cycling is played by rodents, except where permafrost lies near the surface. The ability of rodent populations to consume large quantities of organic matter above and below ground is surprising. For example, in one alpine ecosystem studied in the Rockies, over half the total annual net primary production is consumed by ground squirrels.

13.4 The origin of the Arctic tundra

Clearly only a comparatively short time ago (approximately 8000 years) extensive glaciers covered large areas of the present location of this formation-type. Were there during the last glacial epoch equivalent extensive tundra lands fringing the ice caps? Although characteristic tundra landscape features can be found in former periglacial areas like East Anglia—stone polygons, stone stripes and solifluction deposits, for example—fossil evidence seems to suggest that over most of Eurasia, at any rate, tundra was destroyed and re-evolved several times during the Quaternary (see Figures 6 and 7). During the warmest interglacials, forest seems to have extended well into the Arctic circle itself. In fact, the extensive tundra lands of today seem to be an exception to the pattern of the last two or three interglacials.

13.5 Man and the tundra lands

Low productivity and a harsh climate have inevitably restricted Man's use of these lands. The cultures which evolved in the past were adapted to the ecosystem's severe cycle of abundance and scarcity and are now almost completely degraded by the extension of industrialised cultures. Efforts in Canada and Sweden to offset the breakdown of the ecological pattern of the indigenous peoples by increasing the saleable product from the area have met with only limited success. In Canada, for example, the problems of the domestication of the musk-ox have largely been overcome, although as yet production of its fine wool is practically confined to the experimental station (Poland 1976). In Sweden, the Lapps have turned to exporting reindeer meat and hides, but only where they are able to follow their traditional way of life. However, in all these Arctic lands the greatest threat is not from lack of product but from the demand for labour for mines. It is an accident of geography that these lands coincide with richly-mineralised rocks, and mines and smelters with their needs of labour and road, rail and pipeline communications open up large tracts of country to poorly controlled development and tourist activity. Such difficulties are not confined to the western world. In the USSR the release of large areas of State nature reserve to other uses from 1951 onwards has had considerable impact on the natural ecosystems with the influx of urban populations serving the mines.

References and Further Reading

A book which summarises a great deal of our present knowledge of Arctic and alpine lands is:

**Ivens, J. D. and R. G. Barry 1974. *Arctic and alpine environments.* London: Methuen.

The northern high latitude lands are well dealt with in:

Fuller, W. A. and P. G. Kevan (eds) 1970. *Productivity and conservation in northern circumpolar lands.* Morges (Switz.): Int. Union for the Conserv. of Nature. (This IBP symposium document is much better illustrated than is sometimes the case.)

The Antarctic region is well documented in:

Holdgate, M. W. (ed.) 1970. *Antarctic ecology,* 2 vols. New York and London: Academic Press.

Specific aspects

Billings, W. D. and L. C. Bliss 1956. A comparison of plant development in environments of Arctic and alpine tundras. *Ecol.* **26,** 303-37.

Grubb, P. J. 1974. Factors controlling the distribution of forest types in tropical mountains. In Flenley, J. R. (ed.) *Altitudinal zonation of for. in Malesia.* Hull: Dept. of geography, Hull University.

Mooney, H. A. and W. D. Billings 1961. Comparative physiological ecology of Arctic and alpine populations of Oxyria digyna. *Ecol. Mono.* **31,** 1-25.

Poland, F. 1976. The return of the native. *New Sci.* **69**(992), 616-17.

Steenis, C. G. G. J. van 1968. Frost in the tropics. In *Proc. Symp. rec. adv. trop. ecol.,* Varanasi (1), 154-67.

Terjung, W. H., R. N. Kickert, C. L. Potter and S. W. Swarts 1969. Energy and moisture balance of alpine tundra in mid-July. *Arctic and alpine research* **1**(4), 26-42.

The vegetation of estuaries and sea-shores

The zone between the sea and the land is occupied by a diversity of ecosystem-types developed over a variety of habitats—dunes, brackish lagoons, intertidal mud and sand flats—presented by the coastal zone. In spite of the difficulties presented by salinity, the inshore marine and estuarine habitats are some of the 'most naturally fertile in the world' (Odum 1963). There are several reasons for this:

(a) the rapid circulation of nutrients and food promoted by tidal action;
(b) the trapping of nutrients both physically in mud-derived land and biologically by direct extraction from continually stirred seawater, a process in which both animals and plants participate;
(c) the often more equable local climatic conditions;
(d) the close contact between stages in the food chain.

There is not sufficient space in this book to describe even in outline all the varieties of coastal vegetation-type, but the student will find a good account in Chapman (1976). The present account is confined to a brief examination of the implications of Man's use of the ecosystems, as they are a model for the increasing conflict between short-term need and long-term conservation which, as noted throughout Part Two of this book, afflicts almost every vegetation-type.

The functions of coastal vegetation around the world in relation to Man are various and in many places of great importance. We can summarise the most important aspects as follows:

(i) *The extension of the land surface which may be used for agriculture after suitable treatment*

This process is particularly important in some of the world's great deltas—the Ganges, the Amazon, the Irrawaddy, the Mekong and so on. The mud that is trapped is usually nutrient-rich and in tropical and sub-tropical lands it produces some remarkably high figures for net primary production. For example, in South Vietnam net primary production with new 'high-yield' varieties of rice reaches $20t/ha^{-1}/an^{-1}$, and in Java coastal sugar-cane plantations on former mangrove swamps have one of the highest net primary productions ever recorded at $94t/ha^{-1}/an^{-1}$, which is very near the theoretical potential photosynthetic limits (see Figure 13).

In the tropical world generally, the most important agents in this extension of the land are the mangroves (Figure 56). Although not entirely confined to the tropics—they occur in the Gulf of Aqaba, South Africa and New Zealand—the twenty species or so which make up this highly adapted vegetation spread throughout the tropical world on low, muddy shores and are extraordinarily effective in extending the land surface. Their effectiveness is enhanced by the differentiation of niche amongst the species, some preferring sandy substrates, some muddy conditions, and by the degree of salinity they can tolerate. All possess a form of specialised breathing apparatus (pneumatophores) and can develop stilt roots. They are usually not present where strong wave action is likely.

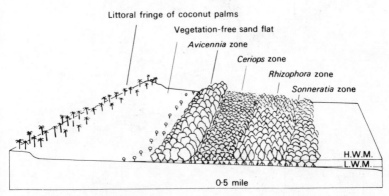

Figure 56. Zonation in mangrove swamps near Tanga, East Africa (data from Walter (1971)). Note that all mangrove species shown have stilt roots.

In *Sonneratia*, for example, the pneumatophores are vertical post-like organs containing fine pores which allow the passage of air at low tide but not water at high tide. Some, for example the red mangrove (*Rhizophora mangle*), have a form of viviparous reproduction, the primary root of the seedling bursting through the hanging fruit as a slender dart-like structure while still carried on the parent tree. The

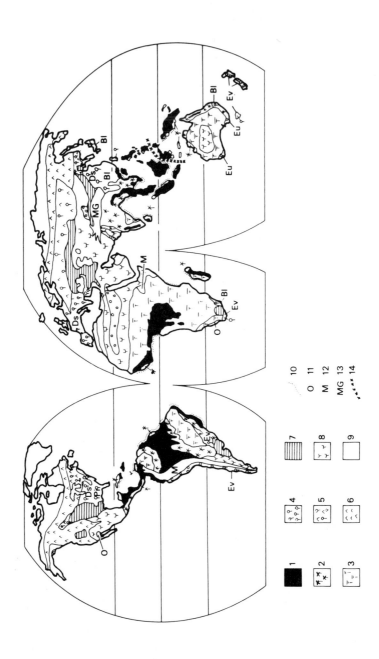

seedling may drop from the parent plant, directly penetrating the mud, and produce lateral roots in a matter of hours, or, if shed at high tide, may be transported to a favourable location. Noisome though it may be, the mangrove swamp is a highly effective ally of man, and where the mangrove forms tall forest on the landward side, it can be a valuable source of timber. In Malaya, for example, regular felling of mangrove forest on a thirty-year rotation long provided construction timber, so much so that there is little climax left. However, mangrove can be easily damaged by pollution, especially those species which prefer brackish water. This seems to be happening in coastal Florida and as industrialisation proceeds along coasts and river basins the mangrove vegetation may be in great need of conservation.

In temperate parts of the world the function of stabilising the muddy coasts is carried out by many species of salt-marsh herbs. These too can be highly effective in increasing the land surface, and the frequent muddy creeks and tidal pools in both temperate and tropical lands have an important role to play in inshore fisheries as they often form the nursery for fish which are later harvested at sea.

(ii) *The protection of lowland from the effects of storm damage*

This role is particularly fulfilled by sand dunes stabilised initially by grasses. Most of these, like the marram grass, are highly xeromorphic and adapted in their rooting to take advantage of fresh sand by springing strong lateral roots into the sand as it is built up. In the succession on dunes they may be followed by xeromorphic woody plants

---

Figure 57. Generalised distribution of the world's vegetation types (from various sources). (1) tropical rain forest; (2) tropical seasonal forest of various kinds; (3) savannas (classified in various vegetation maps into many sub-categories, e.g. thorn forest, thorn woodland and scrub, broad-leaved tree savanna, etc.); (4) forest of warm and cool temperate affinities including deciduous summer forest (Ds), evergreen mixed forest (Ev), broad-leaved evergreen forest (Bl), Eucalyptus forest (Eu) and southern pine forest of USA (Pi); (5) mixed boreal and deciduous forest of the northern hemisphere; (6) boreal forest (also including west coast forest of USA); (7) temperate grasslands; (8) vegetation of arid and semi-arid lands (highly variable, including cactus scrub, thorn scrub and many woody and sparse grass formations); (9) tundra; (10) mangrove coasts; (11) true desert with intermittent plant cover; (12) montane vegetation of the Andes and Abyssinian highlands which includes major formations of tropical montane forest; (13) alpine vegetation of central Asia with extensive grassland areas; (14) Wallace's line.

which can take advantage of the nutrients bonded into the humus created by the dead remains of the grasses. In many temperate lands Man has taken advantage of the stabilised sandy habitats and planted pines, for example in the Landes of France and along the western coast of Jutland in Denmark.

The protective value of the dunes is incalculable. For example, along the southern North Sea from northern Denmark to Cap Gris Nez, they protect densely populated lowland on former salt marshes behind them. When this barrier has been penetrated, as in the great flooding of the twelfth century which created the Zuider Zee, and again in 1953, its natural protection has had to be replaced by a series of artificial barriers constructed at great cost.

The demand for usage of the coastline that increases its commercial value is likely to damage the very ecosystem that protects us at so little expense. In the USA, for example, the eastern seaboard marshes from New Jersey southwards are the scene of a fierce struggle between developers and conservationists, not only over development, but also over coastal pollution. In Europe, the dune coasts with their sandy beaches are under pressure, particularly from leisure demands. Trampling by children can do tremendous damage to the delicate balance of growth on dune land. Although in Britain most local authorities are aware of this damage and attempt to protect dunes, the pressure for access is always there. The estuaries which are the nurseries of so much valuable fish are almost everywhere suffering from pollution, much of which could be avoided, given sufficient political determination by private industry and public bodies to face the costs. As Odum says, 'Conversion of already useful estuaries into open sewers for industrial wastes or into corn fields or house sites for which the topography is not well suited, is not in the best interests of man.'

Coastal vegetation is most certainly one case where the estimate of economic benefits derived from the ecosystem can be assessed in terms for once acceptable to the cost-benefit analyst. Unfortunately, the exercise is rarely attempted and there are few models available to do this. It is significant, for example, that the proposal to use Foulness Island for London's third airport still does not include an estimate of the ecological effects in cost-benefit terms.

References and Further Reading

There are several excellent accounts of mangrove vegetation, and these are reviewed in:

Chapman, V. J. 1973. Mangrove vegetation. In *Proc. pre-congress conf. Bogor of the Pacific sci. assoc.* 1971. (This is not particularly easy to obtain, so the reader is advised to consult the texts below.)

See also:

*Richards, P. W. 1952. *The tropical rain forest.* Cambridge: Cambridge U. Press.

**Walter, H. 1971. *The ecology of tropical and sub-tropical vegetation.* Edinburgh: Oliver & Boyd.

A good introductory account of non-mangrove shore plant biology is to be found in:

*Yonge, C. M. 1949. *The sea shore* (New Naturalist series). London: Collins.

See also:

Stephenson, T. A. and A. Stephenson 1949. The universal features of zonation between tide marks on rocky coasts. *J. Ecol.* **37, 289-305.

Good examples of food-web systems in beaches and estuaries can be found in:

Odum, E. P. 1963. *Ecology.* New York: Holt, Rinehart & Winston.

A good brief summary of dune environments is provided in:

Oosting, H. J. and W. D. Billings 1942. Factors affecting vegetational zonation in coastal dunes. *Ecol.* **23,** 131-43.

Conclusion

Throughout the second part of this book the reader will have noted that in practically every major ecosystem-type from the tundra to the tropical rain forests, the addition of new knowledge is increasing the ability of Mankind to bring vast tracts of the Earth's surface into a managed rather than wild state. The progress of conservation of our great natural heritage of plants and soils is slow and uncertain, but it is undoubtedly one of the most important tasks (if not the most important) of the next few decades. With the world's population expected to double by the first decade of the next century, the need for the basic ecological and biogeographical information on which to base rational policies is urgent. The author hopes that the information provided in this book will enable its readers to participate more effectively in the process of policy-making, either as future conservationists themselves or as informed persons without whom no rational policies can succeed.

Further Reading

Dassman, R. F. 1976. *The conservation alternative.* New York: Wiley. (A hard-headed appraisal of the policies which will be needed in the immediate future by one of America's foremost ecologists.)

Dassman, R. F., J. P. Milton and P. H. Freeman 1973. *Ecological principles for economic development.* New York: Wiley.

Scientific American 1976. Special issue, *Food and agriculture,* **235**(3).

Simmons, I. G. 1974. *The ecology of natural resources.* London: Edward Arnold. (A well-organised appraisal of the world's biotic and other resources.)

Glossary

anaerobic   Without free oxygen.

anion   Negatively-charged particle (non-metallic) in solution.

association   Part of a plant formation dominated by a characteristic species.

autotroph   A plant able to feed itself, i.e. provide its own energy supply.

biomass   The weight of living organisms in an ecosystem. Can be measured as living (fresh) weight, dry weight, organic matter (i.e. ash-free) weight, weight of carbon, units of heat and in various other ways.

biosphere   That part of the Earth in which ecosystems can operate.

catena   Chain of soil types from top to bottom of a slope.

cation   Positively-charged particle (metals and hydrogen) in solution.

clay-humus complex   The electrically-linked mass of clay and humus (colloidal) molecules in soil.

colloids   Substances present in solution where molecules group together to form solute particles. The substances can exist in a mobile (sol) state or in a stable state (gel).

community   The plants and animals which compose the living part of an ecosystem.

diurnal   Relating to daily events.

dominant   A plant which because of its coverage above ground or the root space it occupies below ground controls or modifies the immediate environment and determines the conditions of life for other members of the community.

ecosystem   The complex of the living community and the non-living environment which it inhabits functioning as an integrated system.

ecotone   A zone of transition between two distinct plant communities.

ecotype   A group of local populations within a species from the same type of environment genetically specialised for that environment. (Also called **ecological races.**)

191

eluviation   Process of removal of certain fractions of the clay-humus complex from a soil **horizon.**

endemic   A plant (or animal) restricted to a relatively small geographic area or habitat.

epiphyte   A plant which uses another plant for physical support but without roots in the ground. Epiphytes take no water or other nutrients from the supporting plant.

eutrophic   Well-supplied with nutrients.

evapo-transpiration   The combined loss of water to the atmosphere from a land surface and its plant cover.

flora   All the kinds of plants which make up the vegetation of an area.

formation   A distinct part of a formation-type identifiable geographically.

formation-type   A world vegetation type recognised by its dominance overall by plants of the same life-form.

guttation   The exudation of water as a liquid by plant leaves.

heterotroph   Plant (or animal) which utilises food created by **autotrophs.**

holocoenotic   Wholly unified. Referring to ecosystem components both living and non-living.

horizon   Distinct soil layer.

illuviation   Process of re-deposition in a soil of colloidal material derived from an eluviated horizon or brought by capillary action from depth.

leaching   The removal of dissolved or suspended materials by water percolating downwards.

life-form   A distinctive set of shape, size and other characteristics of morphology shared by a group of related or unrelated species.

limiting factor   Any factor which checks potential growth in an ecosystem.

mutation   A change in the inheritable characteristics of a living thing.

mycorrhizae   Fungi inhabiting the roots of vascular plants.

niche   The role an organism plays in an ecosystem.

nutrient   Any element or compound essential to life.

oligotrophic   Low in nutrients.

photoperiod   Length of daytime in which light is available.

primary production Total amount of organic matter including that used by plant respiration during a period of measurement. Also called **gross primary production.** Net primary production is the organic matter stored in plant tissues in excess of respiration during a period of measurement. In relation to vegetation both figures are usually given as mean weights per unit area.

profile The arrangement of soil horizons from surface to bedrock.

radiation Emission and transmission of energy in wave form, applied to both electromagnetic energy and particles from radioactive substances.

sere A series of plant communities which successively occupy the same piece of ground.

synusia A set of plants within a complex community which share the same life-form.

transpiration The process of water-vapour release to the atmosphere from the above-ground parts of plants.

valency Combining power of an atom (measured as the number of hydrogen atoms which an atom will combine with or replace).

zonal vegetation A broadly conceived vegetation unit which occupies a larger part of the surface in areas without significant human ecological effects and is dominated by the macroclimatic characteristics. (Azonal vegetation is that dominated by extreme soil characteristics.)

Index